Information Security Governance using Artificial Intelligence of Things in Smart Environments

This book explores the integration of Artificial Intelligence (AI) with the Internet of Things (IoT) to address security challenges in smart environments. It delves into how AI enhances the governance of information security by automating processes, detecting threats, and ensuring the protection of data in interconnected IoT systems. It covers theoretical foundations, practical frameworks, and case studies, offering insights into securing smart cities, homes, industries, and healthcare systems. It also emphasizes governance models that leverage AI to manage security policies and risk in dynamic, data-driven ecosystems.

This title focuses on the study and application of AI of Things in the field of information security governance. Intelligent environments, characterized by increasing connectivity of devices and systems, present unique challenges for information security. The use of AI of Things offers opportunities to enhance security in these complex environments.

Information Security Governance using Artificial Intelligence of Things in Smart Environments

Edited by

Mariyam Ouaissa, Mariya Ouaissa, Tarik Hidar, Ram Chandra Sachan, Akhil Mittal, and Sanjay Poddar

CRC Press
Taylor & Francis Group
Boca Raton London New York

CRC Press is an imprint of the
Taylor & Francis Group, an **informa** business

Designed cover image: Shutterstock Image ID 2483457133

First edition published 2026

by CRC Press
2385 NW Executive Center Drive, Suite 320, Boca Raton FL 33431

and by CRC Press
4 Park Square, Milton Park, Abingdon, Oxon, OX14 4RN

CRC Press is an imprint of Taylor & Francis Group, LLC

ISBN: 978-1-032-99814-5 (hbk)
ISBN: 978-1-032-99850-3 (pbk)
ISBN: 978-1-003-60630-7 (ebk)

DOI: 10.1201/9781003606307

Typeset in Sabon
by SPi Technologies India Pvt Ltd (Straive)

Contents

Author/editor biographies

Mariyam Ouaissa is currently an Assistant Professor in Networks and Systems at ENSA, Chouaib Doukkali University, El Jadida, Morocco. She received her Ph.D. degree in 2019 from National Graduate School of Arts and Crafts, Meknes, Morocco and her Engineering Degree in 2013 from the National School of Applied Sciences, Khouribga, Morocco. She is a communication and networking researcher and practitioner with industry and academic experience. Dr. Ouaissa's research is multidisciplinary that focuses on Internet of Things, M2M, WSN, vehicular communications and cellular networks, security networks, congestion overload problems, and resource allocation management and access control. She serves as a reviewer for international journals and conferences including *IEEE Access*, *Wireless Communications* and *Mobile Computing*. Since 2020, she is a member of the International Association of Engineers (IAENG) and International Association of Online Engineering, and since 2021, she has been an ACM Professional Member. She has published more than 60 research papers (this includes book chapters, peer-reviewed journal articles, and peer-reviewed conference manuscripts), 15 edited books, and 6 special issues as guest editor. She has served on program committees and organizing committees of several conferences and events and has organized many symposiums, workshops, and conferences.

Mariya Ouaissa is currently a Professor in Cybersecurity and Networks at Faculty of Sciences Semlalia, Cadi Ayyad University, Marrakech, Morocco. She holds a Ph.D. degree, having graduated in 2019 in Computer Science and Networks from ENSAM-Moulay Ismail University, Meknes, Morocco. She is a Networks and Telecoms Engineer, graduated in 2013 from National School of Applied Sciences in Khouribga, Morocco. She is a Co-Founder and IT Consultant at IT Support and Consulting Center. She worked for the School of Technology of Meknes Morocco as a Visiting Professor from 2013 to 2021. She is member of the International Association of Engineers and International Association of Online Engineering, and since 2021, she is also an ACM Professional Member. She is Expert Reviewer with Academic Exchange Information

Centre (AEIC) and Brand Ambassador with Bentham Science. She has served and continues to serve on technical program and organizer committees of several conferences and events and has organized many symposiums, workshops, and conferences as a General Chair. She has also been a reviewer of numerous international journals. Dr. Ouaissa has made contributions in the fields of information security and privacy, Internet of Things security, and wireless and constrained networks security. Her main research topics are IoT, M2M, D2D, WSN, cellular networks, and vehicular networks. She has published over 70 papers (book chapters, international journals, and conferences/workshops), 20 edited books, and 10 special issues as guest editor.

Tarik Hidar is currently a Professor in Cybersecurity at Private University of Marrakech (UPM), Marrakech, Morocco. He is a seasoned expert in the fields of cybersecurity, network, and telecommunications engineering. His career began with a solid foundation in engineering, where he gained extensive experience in designing and optimizing secure and efficient communication infrastructures. As a network and telecommunications engineer, he delved into the complexities of data transmission, protocol management, and network security, which sparked his passion for cybersecurity. To further his expertise, he pursued a Ph.D. in cybersecurity, focusing on the emerging field of the tactile internet. His doctoral research explored the unique security challenges of this next-generation technology, which aims to provide ultra-low-latency communication for critical applications such as remote surgery, autonomous vehicles, and augmented reality. He developed innovative security frameworks and protocols to safeguard data integrity, confidentiality, and availability in these high-stakes environments. In addition to his technical and research expertise, he is also an expert trainer in various domains, including computer networks, system administration, cybersecurity, intrusion testing, digital forensics, artificial intelligence (AI), and big data. He has had the opportunity to work with several esteemed training organizations, such as M2i, Global Knowledge, Docapost, Demos, and AB Conseils. Throughout his career, he has remained committed to bridging the gap between cutting-edge technological advancements and the security measures required to protect them. Today, he continues to be actively involved in research, development, and training, ensuring that the next generation of technologies is both innovative and secure.

Ram Chandra Sachan is a CCIE Wireless (#60766) with over 22 years of experience in network design, architecture, implementation, and support. He is currently working as a Network Architect with Wipro, based out of Greensboro, NC. As an active member of IEEE and the IEEE Communications Society, he has been involved in various capacities, including serving as an organizing/program committee member for

ICFNDS 2024 and CCSN 2024. Additionally, he has reviewed papers for ICBDS 2024 and has served as a TPC member for CIoT 2024, LATINCOM 2024, and ICACTCE 2024.

Akhil Mittal is a cybersecurity thought leader with 20+ years of extensive experience in application security, cloud security, DevSecOps, and enterprise security programs. He has led strategic initiatives for major clients, specializing in vulnerability assessments, penetration testing, and secure software development lifecycles. As a Gartner Cybersecurity Ambassador, Akhil shares insights on emerging cybersecurity trends, including AI-driven attacks and software supply chain risks. His commentary is regularly featured in top media outlets such as SC Media and Dark Reading, where he offers practical advice and unique perspectives on critical security challenges.

Sanjay Poddar brings a unique perspective to editing, combining his extensive IT background with a passion for clear communication. With expertise spanning networks, network security, and technology solutions, Sanjay offers invaluable insights into technical writing and documentation. His experience across the news and media, telecom, and cybersecurity industries allows him to bridge complex technological concepts with accessible prose. Sanjay's keen eye for detail, honed through years of project management, ensures precision in every manuscript he touches. His collaborative approach and leadership skills foster productive relationships with authors, helping them articulate their ideas with clarity and impact. Always attuned to emerging trends, Sanjay's forward-thinking editorial style keeps publications at the cutting edge of technological discourse.

Preface

The book explores the integration of Artificial Intelligence (AI) with the Internet of Things (IoT) to address security challenges in smart environments. It delves into how AI enhances the governance of information security by automating processes, detecting threats, and ensuring the protection of data in interconnected IoT systems. It covers theoretical foundations, practical frameworks, and case studies, offering insights into securing smart cities, homes, industries, and healthcare systems. It emphasizes governance models that leverage AI to manage security policies and risk in dynamic, data-driven ecosystems.

The book focuses on the study and application of AI of Things in the field of information security governance. Intelligent environments, characterized by increasing connectivity of devices and systems, present unique challenges for information security. The use of AI of Things offers opportunities to enhance security in these complex environments.

The main objective of this book is to explore how AI of Things can be integrated into information security governance practices to improve data protection, threat detection, and incident response. This involves examining various aspects, such as risk analysis, threat detection, and incident prevention and response.

This book addresses a crucial area as intelligent environments become increasingly prevalent. It aims to include case studies or simulations of real implementations of AI of Things in intelligent environments to illustrate its effectiveness and challenges. In addition, it provides practical guidelines and recommendations for the successful integration of AI of Things into information security governance, in order to strengthen data protection and system resilience in these complex environments.

Let's take a closer look at the specific themes and contributions of each chapter:

Chapter 1: Starting with foundational concepts establishes the basis for understanding AIoT and security.

Chapter 2: Governance strategies follow naturally, focusing on trust and accountability as key pillars.

Chapter 3: Highlighting threats early on provides a context for the importance of robust governance.
Chapter 4: Discussing governance frameworks after threats connects high-level policies with practical challenges.
Chapter 5: Introducing an advanced threat detection framework ensures a transition to more applied, technical solutions.
Chapter 6: Cloud-integrated IoT systems expand on the technical focus, emphasizing network security.
Chapter 7: Urban safety and AI-driven solutions diversify the application of AIoT in broader contexts.
Chapter 8: Transitioning into healthcare begins a sector-specific focus, outlining challenges and opportunities.
Chapter 9: Following with trust in smart medical devices ties into healthcare systems and trust-building.
Chapter 10: Exploring the role of AIoT in healthcare ecosystems, unique security concerns, and AI-driven solutions.
Chapter 11: Concluding the healthcare section with describing how AI and IoT technologies enhance healthcare systems by providing accurate diagnostics, personalized treatments, and real-time health tracking.
Chapter 12: Shifting to industrial IoT (IIoT) introduces a new domain of application.
Chapter 13: Ending with optimized IIoT security ensures the book concludes with applied, advanced security strategies.

Contributors

Abderrahim Abdellaoui
Ibn Tofail University
Morocco

Qasem Abu Al-Haija
Jordan University of Science and Technology (JUST)
Jordan

Shahad AL-Tamimi
Princess Sumaya University for Technology (PSUT)
Jordan

Danish Ali
Wuhan University
China

Md Shoeb Atthar
Methodist College of Engineering and Technology
India

Vasavi Sravanthi Balusa
Methodist College of Engineering and Technology
India

Kishor Kumar Reddy C
Stanley College of Engineering and Technology for Women
India

Srinath Doss
Botho University
Botswana

Areesha Fatima
Stanley College of Engineering and Technology for Women
India

Meenal H
Methodist College of Engineering and Technology
India

Marlia Mohd Hanafiah
Universiti Kebangsaan Malaysia
Malaysia

Pagadala Indira
Keshav Memorial Institute of Technology
India

Sundas Iqbal
Nanjing University of Information Science and Technology (NUIST)
China

Khawla Jabari
Ibn Tofail University
Morocco

Nikitha Jillapalli
Methodist College of Engineering and Technology
India

Sandeep Kampa
Splunk-Cisco
USA

Fahad Rashad Khan
University of Haripur
Pakistan

Harika Koormala
Methodist College of Engineering and Technology
India

Srinivas Reddy Kosna
Splunk-Cisco
USA

Kari Lippert
University of South Alabama
South Alabama, USA

Avula Mahathi
Stanley College of Engineering and Technology for Women
India

Deepika Malve
Keshav Memorial Institute of Science and Technology
India

Hatem Mosa
Princess Sumaya University for Technology
Jordan

Hind Moussaid
Ibn Tofail University
Morocco

Sainag Nethala
Splunk-Cisco
USA

Mariya Ouaissa
Cadi Ayyad University
Morocco

Mariyam Ouaissa
Chouaib Doukkali University
Morocco

Sanjay Poddar
Fortinet
USA

Sumaira Rafique
National University of Computer and Emerging Sciences (Sub campus Milad street, Faisal Town, Lahore)
Pakistan

Ch Rajyalakshmi
Methodist College of Engineering and Technology
India

Ram Chandra Sachan
Wipro
USA

Wasswa Shafik
School of Digital Science, Universiti Brunei Darussalam, BE1410, Gadong, Brunei-Dig Connectivity Research Laboratory (DCRLab), 600040
Uganda

Thakur Monika Singh
Stanley College of Engineering and Technology for Women
India

Chapter 1

Foundations of artificial intelligence and IoT in security

A pillar for modern security governance

Avula Mahathi, Kishor Kumar Reddy C, Thakur Monika Singh, and Srinath Doss

1.1 INTRODUCTION

The sophistication of the current generation of smart technologies presents the security domain with unique new dimensions. As globalization in the communication system has continued, the use of interconnected systems and devices is a phenomenon that has exposed the weakness of previous security systems. The new trend of the contemporary world is the combination of artificial intelligence (AI) and the internet of things (IoT), which provides improved security with intensified tools and approaches [1]. Not only do these technologies help give a real-time view, but it also offers predictive qualities to security systems, improving its performances. This chapter introduces AI and IoT and more specifically their convergence in security, as a way of solving the current issues experienced in smart environments. They both explore the use of governance in achieving secure and ethically sound improvements, thus offering coverage of the expanding contours.

1.1.1 Overview of AI and IoT in security

While AI and IoT are two completely different categories of technologies, their application in security has synergies. AI equips systems with the characteristic of analyzing large volumes of data while learning from them and making predictions about possible risks. Some of these include machine learning to identify the outliers, natural language processing to understand language as well as computer vision in making decisions. However, IoT acts as a platform of connectivity where a large number of connected devices and sensors are integrated to generate a meshing network. IoT devices gather and forward fresh data, and an active stream of information is then provided to the AI algorithms which then analyze data to detect and prevent risks. For instance, in a smart city context, in objects such as cars and roadways, weather and light sensors, and power consumption meters, IoT and AI are used in smart city detection and response to abnormal activities. Both AI and IoT complement each other to ensure simple device security as well as enhance the requisite network security against enhanced cyber threats [2].

DOI: 10.1201/9781003606307-1

1.1.2 Importance of security governance in smart environments

With the increase in smart environments, there is a concern on the interconnectivity that such environments bring about. Threats that are associated with cyberspace and information technology are comprehensive since they may affect safety of the public, basic infrastructures, and personal privacy. Risk management is the process of identifying and responding to risks in order to minimize the damage they cause, and security governance provides a structured approach to managing the aforementioned risks through policy, standards, and procedures.

AI and IoT also highly improve the governance outcomes through the possibility of conducting operations and responding to incidents quickly [3]. For example, AI detection algorithms can detect a set parameter that points to a possible hack and counteract it before it can occur, whereas IoT offer a constant chance to monitor an environment and gather evidence in case of a breach. Security governance thus makes sure that these technologies are implemented appropriately with an equal regard to the ethics besides ensuring that they have complied with legal requirements. This is especially something that needs to be done in the special industry such as medical facilities and financial realms where the enterprise's risks are higher.

1.1.3 Scope and objectives of the chapter

This chapter will focus on AI and IoT's basic concepts and applications in security. The purpose is to focus on their positive impact and discuss the issues regarding their usage. Key objectives include:

Understanding Core Concepts: Explain AI and IoT definitions, their primary concepts that connect to the security topic.

Exploring Real-World Applications: Explain basic ideas of IoT, AIoT, and several cases in industries such as smart city, medical, and manufacturing.

Identifying Challenges: Discuss matters such as security concerns, compatibility, and application of AI and IoT technologies as moral questions.

Emphasizing Governance: Stress on the value of requirement legislatives and norms in guaranteeing safe and asymptotic artificial intelligence of things.

Envisioning Future Trends: Providing the discussion of novel technologies and tendencies in governing the AIoT security in the future.

Table 1.1 indicates most of the major dissimilarities between traditional security systems and AIoT-based security systems in significant operational as well as performance attributes. Incorporating traditional security

Table 1.1 Comparison of traditional security systems vs. AIoT security systems

Aspect	*Traditional security systems*	*AIoT security systems*
Response Time	Manual or delayed responses due to human intervention.	Real-time detection and automated threat response.
Scalability	Limited to specific infrastructure or setups.	Easily scalable to accommodate expanding IoT networks.
Threat Detection	Reactive, identifying threats after breaches occur.	Proactive, predicting, and mitigating threats before they materialize.
Data Analysis	Static and limited analysis capabilities.	Advanced, real-time data analytics using AI algorithms.
Adaptability	Struggles with dynamic, evolving threats.	Adaptive, learning from new threats and evolving accordingly.
Cost-Effectiveness	Higher operational costs for manual oversight.	Reduced costs through automation and efficiency.
Interconnectivity	Limited device and system integration.	Seamless integration across diverse IoT devices and platforms.

measures into an organization creates opportunities for vulnerabilities because it is basically a rigid and time-consuming approach to security. These systems are usually post-breach systems as they identify threats after some unauthorized access, and they cannot change course to address new threats. On the other hand, AIoT security systems use technologies such as artificial intelligence (AI) and the internet of things (IoT) to provide preventive and to some extent self-actuated solutions. Predictability, detectability, and preventability are achievable through real-time monitoring and harnessing of big data analytics by the AIoT systems. They provide simple and efficient solutions that can potentially accommodate increasing numbers of IoTs and achieve proper connectivity between them. Moreover, cost-optimization results from automation, for which substantial manual intervention is not required. This comparison shows how AIoT shifts the classic security approach by adding flexibility, effectiveness, and the ability to forecast into modern security models to accommodate the emerging and continuous dynamic security threats in current complex environments.

1.2 FOUNDATIONAL CONCEPTS OF AI AND IoT

1.2.1 Key features of artificial intelligence in security

AI in particular has come a long way and is now at the inflexion point of jump starting security systems in several domains. Its capacity to analyze big data, capture patterns, and make informed decisions has enhanced traditional ways of security. Some of the key features of AI in the context of security are:

Anomaly Detection: Machine learning algorithms are very effective in detecting anomalies, which more often than not suggest incidents of insecurity. For instance, with AI, one can easily identify an irregular login location or access or atypical behavior of the system before an attack amps up.

Predictive Threat Modeling: AI makes it easy to guard against threats through the analysis of past events with a view to predicting future attacks. AI can also learn that a specific vulnerability can cause cyber threats, and as a result, can cause preventative measures to be taken with reduced risk of the threats being successful.

Automation of Security Tasks: Log analysis and vulnerability scanning tasks that are repetitive and require large amounts of time to complete are performed with the help of AI. It also makes security operations more accurate and quicker while experts can dedicate time to analyzing more complicated situations.

Just like in client interfacing and application, NLP is also used in Threat Intelligence.

NLP allows the use of AI to analyze text and this encompasses security reports, emails, and many other forms of data. This aids in detecting current phishing scams, fake news, or any other that is being, or has been, mentioned on the World Wide Web or social platforms.

Real-Time Decision-Making: AI offers the tangible power to make decisions in the blink of an eye with the aid of real-time data – data which is up-to-date. For example, an AI in smart surroundings may know and prevent the connection of an unauthorized user, quarantining infected devices, or redirecting traffic in a cyberattack, with marginal downtime.

Adaptive Education: Over time, AI systems' performance improves as they continue to learn and develop. By changing their models in response to emerging threats, they make sure that security frameworks continue to be strong even as attackers create increasingly complex techniques.

AI is essential to contemporary security frameworks because of these characteristics [3]. How businesses protect their digital and physical assets is changing as a result of its capacity to support human activities while offering scalable, accurate, and proactive solutions. We will examine how these features work with IoT to develop intelligent and robust security systems in the sections that follow.

Table 1.2 Key features of AI and IoT technologies

Feature	*Artificial intelligence (AI)*	*Internet of things (IoT)*	*Application in security*
Core Components	Machine learning, neural networks, natural language processing.	Sensors, actuators, communication protocols (e.g., Wi-Fi, Zigbee).	Threat detection, anomaly detection, and event prediction.
Data Processing	Processes large datasets to identify patterns and insights.	Gather real-time data from interconnected devices.	Real-time analysis for proactive threat management.
Decision-Making	Autonomous and adaptive decision-making capabilities.	Relays actionable data to central systems or AI modules.	Enables intelligent responses to detected threats.
Scalability	Highly adaptable across different applications.	Expands seamlessly with network growth.	Provides scalable security solutions for large systems.
Primary Strength	Analytical and predictive power.	Physical connectivity and data acquisition.	Together, enable advanced, proactive security systems.

Table 1.2 simply shows how AI and IoT are balanced and can complement each other in security. AI is able to facilitate analytical features for data processing, while IoT stands for the source of actual data coming from physical objects. Combined, these technologies increase the speed, accuracy, and flexibility of current security systems at large.

1.2.2 Role of IoT in building secure ecosystems

A strong focus on IoT systems is mandatory for generating secure ecosystems because IoT allows connections and real-time control of numerous devices. In a smart environment, things that are capable of IoT [4], including sensors, cameras, and smart locks, are constantly collecting data and transmitting data to offer real-time security. These devices increase awareness to prevent threats by giving an opportunity to security teams to see and control activities in real time. For instance, in smart buildings, the IoT sensors are capable of sensing some odd motion or shifts in the environment and this raises an alarm to the security.

It also includes management from distance that is among the most crucial for sectors such as healthcare or industrial automation for example. Internet of things (IoT) devices can transmit information to centralized systems for assessment and to see if security measures have been violated or threats detected warrant remote control [5]. Furthermore, IoT makes quicker and easier security procedures since if IoT sensors detect the breach of security they can themselves immediately shut the doors, close the access points, or notify the user/administrator.

However, with the growth of IoT, user-based security is also constrained by some of the more security problems such as unauthorized access, weak authentication, and vulnerability of devices. These challenges require that the technology used in IoT be enhanced to higher forms such as AI technology.

1.2.3 AIoT: The synergistic integration of AI and IoT

AIoT stands for Artificial Intelligence of Things through which AI is combined with IoT to lead to enhanced smart, intuitive, and self-learning security solutions. This integration improves on the efficiency of both technologies by improving their security capabilities [6]. AI provides smart decision-making to the IoT network though otherwise non-intelligent IoT devices are merely used for collection and transmission of data. Applying machine learning results in AI being able to consider huge amounts of data collected from IoT devices and analyze them for potential threats and potential actions in order to prevent those threats. For instance, in a smart factory, other IoT devices are used to track and record the performance of certain apparatus, and AI is able to read these data to identify evidence of a breakdown or impending failure. When a device behaves paradoxically, AI can respond proactively and quickly such as calling the attention of personnel/staff, or triggering the maintenance process so that forced downtime and threats can be avoided. Likewise, AIoT can identify new trends of abusive use of the network or attempts at unauthorized access in real-time, which can then lead to prompt action including blocking access or activating the lock down mechanism.

It also means that the integration of AI and IoT also leads to proactive security features. By analyzing data, AI can notify IoT devices of some threat and mitigate it before growing into a major security compromise. Moreover, these models are more efficient and adaptive since they update from new data, making the system more useful, the longer it is used and more protected from security threats [7]. Unlike conventional peripheral security approaches, AIoT not only improves the security of individual devices but also provides a big picture of system activities and a real-time defense line against escalating threats. This makes AIoT a success factor in developing safety and solidity in intertwined environments for smart living.

Table 1.3 also represents typical IoT security issues like unauthorized access and data leakage and explains how AI can solve them. With the help of such AI solutions as anomaly detection, encryption, and proactive traffic analysis, the organizations are able to protect IoT devices and networks from new threats.

Table 1.3 Common IoT security risks and AI solutions

IoT security risk	*Description*	*AI solution*
Unauthorized Access	Hackers gaining access to IoT devices and networks.	AI-driven access control and biometric authentication.
Data Breaches	Leakage or theft of sensitive data from IoT devices.	AI-based encryption and secure data transmission.
Device Hijacking	Compromised IoT devices used for malicious purposes.	AI anomaly detection to identify unusual behavior.
DDoS Attacks	Overloading IoT systems with malicious traffic.	AI traffic analysis to detect and block attack patterns.
Lack of Updates	Vulnerabilities due to outdated IoT firmware.	AI-powered patch management to identify and deploy updates.
Interoperability Issues	Inconsistent communication among diverse IoT devices.	AI systems to optimize protocol compatibility and ensure smooth operations.

1.3 APPLICATIONS OF AI AND IoT IN SECURITY

Across many industries, the security landscape has changed as a result of the combination of artificial intelligence (AI) and the internet of things (IoT). Companies may create security systems that are more effective, proactive, and adaptable by utilizing AI's capacity to handle and analyze enormous volumes of data and the IoTs' extensive network of linked devices. Predictive maintenance, automatic reactions to possible risks, anomaly detection, and real-time monitoring are all made possible by these technologies. When AI and IoT are combined, they offer improved capabilities for safeguarding vital infrastructure, identifying weaknesses, and guaranteeing the security of digital and tangible assets. In an increasingly linked world, this collaboration is essential to building safe, robust ecosystems.

1.3.1 Vulnerability detection and threat mitigation

One of the greatest measures of manifestation of AI and IoT in security is their entire capability of identifying risks and responding to them. IoT devices may consist of sensors, cameras and network monitors that actively stream data to the AI systems for analysis, to detect deviation and potential security threats [8]. It involves identifying some anomaly, anomalous behavior or generally anything that appears as out of the ordinary and which may depict a breach into the security system. It is important to note that machine learning algorithms of AI conduct vulnerability detection. For instance in a smart building, AI systems evaluate data gathered from several IoT sensors to determine whether there are motions that are strange, efforts to infiltrate, or changes in climate, for example the temperature or moisture content. AI, therefore, can alert or respond automatically of an existence or a planned pattern deviation and or prompt a human operator for further evaluation. This capability could enable identification of threat ahead of time, thereby preventing its damage bend from effecting itself on the organization.

AI helps improve threat management in another aspect through its ability to respond swiftly and autonomously [9]. For example, if an IoT in the smart city of the real estate detects intrusion, AI systems can immediately stop the potential cause, isolate affected devices or re-route connections to avoid additional infections. Quick and efficient countermeasures exist as AI easily identifies threats and processes data quickly for backup measures, that is, reducing the chances of successful attack. Besides working in real time, learning from new data accumulated over the years revealed AI and IoT systems are unmatched. AI models can learn from past incidents and continuously adjust their algorithms with a view to not falling foul of very smart attacks. Such adaptability enables the AIoT systems operators to constantly implement the best strategy of operating the systems in a competitive digital landscape.

Organizations can create more thorough and sophisticated security solutions that not only identify and neutralize threats but also stop possible breaches from happening by combining AI and IoT [10]. These apps help protect smart environments from a variety of attacks by taking a more proactive and effective approach to cybersecurity. Table 1.4 presents various AIoT solutions for improving security based on the different domains. These case studies also demonstrate how powerful data processors such as AI and connected devices such as IoT can build strong solutions [11]. For instance, in the smart surveillance system, AIoT improves threat identification with facial recognition and suspecting activities while investigating. AIoT systems then scout and protect such critical infrastructures when unauthorized people or suspicious events are noticed.

AIoT in smart cities maintains public security by constantly scrutinizing the environment, and in healthcare, it protects personal information and medical equipment. In the same way, AIoT applications are applied in automobile and finance since their problems are also specifically distinct including the protection of autonomous cars from hackers and fraud in monetary affairs. These examples show that AIoT is quite capable of revolutionizing contemporary security systems.

1.3.2 Real-time monitoring, data analytics, and use cases in critical infrastructure protection

AI and IoT spearhead the provision of security through monitoring and processing of data in real time especially for issues to do with protection of critical infrastructure. Therefore, IoT portals combined with AI provide the constant data acquisition of systems conditions and state and enable the immediate identification of threat incidents and timely containment of the

Table 1.4 AIoT use cases in various security domains

Security domain	*AIoT application*	*Description*	*Benefits*
Smart Surveillance Systems	AI-powered CCTV and IoT-enabled cameras	Combines AI for facial recognition and IoT for real-time streaming and centralized data storage.	Enhanced threat detection and reduced response times.
Critical Infrastructure	Intrusion detection and infrastructure monitoring	AI analyzes sensor data from IoT devices to detect anomalies or unauthorized access.	Prevents disruptions and ensures operational safety.
Smart Cities	Real-time threat monitoring systems	AI processes data from IoT sensors placed in public spaces to identify potential security risks.	Promotes public safety and rapid response to incidents.
Healthcare Security	Securing connected medical devices	AI monitors IoT-enabled medical equipment for signs of tampering or malfunction.	Protects patient data and ensures device reliability.
Automotive Security	Securing autonomous vehicles	AI detects cyber threats targeting IoT components in self-driving cars, such as GPS and sensors.	Ensures passenger safety and prevents system failures.
Financial Sector	Fraud detection in connected devices	AI analyzes transactional data from IoT-enabled ATMs and POS systems to identify fraud patterns.	Reduces financial losses and boosts consumer trust.

situation [12]. Underlying segments of IoT devices include power or electricity grids, transport systems, and water supply systems where IoT devices monitor and relay updated information from multiple data sources: sensors, video surveillance systems, controls, and sensors that measure environmental conditions [13]. This data is related to higher systems for processing and or analysis. To make sense of this massive amount of real-time data, AI algorithms determine patterns, recognize possible risks, and identify symptoms of failure or malign intentions [14]. For instance, IoT sensors track the condition of power components in a power grid; then AI studies this data for signs of failures or attempted cyberattacks, including power variations. If a threat is detected, AI can call for corrective action for instance diverting the power, informing the system operators or even shutting down the area in order to avert further problems.

The implications of big data go beyond threat identification; it may also be used for future planning, for situations where one needed to monitor patterns and indicators, and for scheduling. Through analysis of temporal data, AI systems are able to forecast the time and place of likely occurrence of a vulnerability to enable an organization to take precaution in order to avoid failures or an attack [15]. The analysis results in this capability expressing the ability to forecast problems and breakdowns ahead of time, and thereby decrease job interruptions, improve operations, as well as lower the possibilities of major breakdowns. The combination of IoT with AI guarantees that any anomalies or possible security breaches are identified early, allowing for prompt responses that protect infrastructure and public safety in industries where operational continuity is crucial, such as healthcare, transportation, and industrial control. These solutions also guarantee adherence to industry rules, offering accountability and transparency in the protection of critical systems.

All things considered, the combination of IoT's data gathering skills and AI's analytical prowess provides improved security and operational resilience in critical infrastructure, making it a vital component for safeguarding vital services in a quickly changing digital environment.

1.4 CHALLENGES IN IMPLEMENTING AI AND IoT FOR SECURITY

While there are many advantages to using AI and IoT in security, there are also a number of issues that must be resolved to guarantee the efficiency and long-term viability of these technologies. How well AIoT systems can be implemented, maintained, and expanded across various industries depends on these issues [16]. The subject of AI and IoT integration in security revolves around issues like resource management, scalability, interoperability, and data privacy. Although integrating AI and IoT into security systems has many benefits, there are a number of issues that need to be resolved for implementations to be safe and successful. Concerns about data privacy and ethics, maintaining compatibility among a wide variety of IoT devices, and efficiently managing scale and resources are some of the major obstacles. For AIoT solutions to successfully provide safe, effective, and resilient ecosystems, these obstacles must be overcome.

1.4.1 Data privacy, interoperability, and scalability challenges

Two of the most important issues that are pertinent to the successful integration of AI and IoT for security involve protection of data privacy and the problem of ethics. IoT devices produce massive volumes of personal information, which if processed unsafely, would cause privacy breaches or misuse. For instance, in smart homes, or healthcare, IoT sensors, and devices gather personal information that is highly sensitive. This is why it is important to use technologies such as secure encryption and proper access controls along with legislation such as GDPR to prevent data misuse and protection. Lack of privacy can result in member prosecution, lack of trust and increased risks of hacking. The last major issue in the deployment of AIoT security system is interconnectivity. P vs S: Granularity – This includes some of the barriers, such as differences in protocols and standards among IoT device manufacturers, which make integration challenging [17]. In this setup, if various devices do not have a common protocol of operation, the efficiency of the security system deteriorates. This lack of coordination can converge problematic issues within the system, making the responses to threat slower as well as increasing the risks for the whole security system. Achieving interoperability so that the devices can work together in harmony with common interfaces and protocols remains critical to the establishment of AIoT security frameworks.

Last but not least, two more fundamental requirements are needed when it comes to AIoT security system deployment: scalability and resource control. With the rise in internet connection devices, there is a challenge of dealing with the performance of the system, resource management, dealing with the huge amount of data which requires processing in real time. AI algorithms and IoT systems need large computational resources and storage to process this data, especially for application in big settings like smart cities and other critical infrastructures [18]. If the correct escalation plan and resource investment in strong, generalized architectures are not followed, the application may exhibit poor global response times, data traffic jams, and depleted resources that can compromise the security solution [19].

Meeting these requirements demands a RI approach which has been discussed above such as open standards, data governance, infrastructure preparedness among others. These issues, when solved, will enable organizations to harness the full potential of AI and IoT in delivering the goal of secure and resilient systems [20]. The IoT security needs of AIoT systems along with the implications, challenges and solutions or research domains are stated in Table 1.5. Problems such as data privacy and compatibility problems are found to exist because of the large amount of personal data and numerous IoT applications. These challenges are solved through the usage of Artificial Intelligence applied to encryption, availability of standardized protocols and utilization of adaptive framework. Furthermore, concerns of scalability and resource limitations present the technical implications of IoT

Table 1.5 Key challenges and solutions in AIoT security

Challenge	*Description*	*Proposed solution/ Ongoing research*	*Impact*
Data Privacy Concerns	Unauthorized access to sensitive data collected by IoT devices.	Implementation of AI-driven encryption techniques and decentralized data management.	Ensures robust protection of user data and compliance with regulations.
Interoperability Issues	Lack of seamless communication between heterogeneous IoT devices and platforms.	Development of AI algorithms to enhance protocol standardization and system compatibility.	Promotes cohesive and efficient IoT ecosystems.
Scalability Challenges	Difficulty in managing security for expanding IoT networks.	AI-powered adaptive security frameworks that scale with network growth.	Reduces vulnerability and maintains consistent security across devices.
Resource Constraints	Limited computational power and energy in IoT devices for running security measures.	Lightweight AI models optimized for edge computing and energy-efficient operations.	Enhances security without compromising device performance.
Ethical and Regulatory Gaps	Absence of universal guidelines for ethical AIoT implementation.	Formulation of comprehensive governance policies and adherence to global security standards.	Builds trust and fosters responsible deployment of AIoT systems.
Real-Time Threat Mitigation	Challenges in detecting and responding to threats instantly.	Integration of AI for real-time threat analysis and automated incident response.	Minimizes damage and improves response efficiency.

networks, particularly in large and growing IoT systems in the network. These are managed through light AI designs and edge adaptations. The table also drives the point for ethical and regulatory policies to guide the deployment of AIoT systems across the world [21]. Last but not least, solutions envisaged for real-time threat handling exploit AI's features to address threats when they occur, hence enhancing system robustness.

Such an approach guarantees the optimal implementation of AIoT in security and minimization of threat to privacy, time, and ethics.

1.5 GOVERNANCE AND REGULATORY FRAMEWORKS

To guarantee that AI and IoT be used in a secure and morally responsible manner, they must be effectively governed and regulated. Strong governance frameworks and adherence to pertinent regulatory norms will be essential for data protection, privacy preservation, and system dependability as AI and IoT technologies become more integrated into vital industries. In order to steer organizations toward safe and moral AIoT activities, standards, regulations, and international initiatives are essential.

1.5.1 Standards for AI and IoT security systems

AI and IoT development require the specification of certain norms or guidelines that could be followed by developers in order to make the systems as interoperable, reliable, and secure as possible. These aid in establishing optimal practices with regard to the protection of devices that are used in the implementation of an organization's activities, shielding of data, and incorporation of a system into an organization. Some global and specific industrial trends involve development of standards for the fast and secure deployment of secure AI and IoT systems. For instance, International Organization for Standardization (ISO) and Institute of Electrical and Electronics Engineers (IEEE) are the primary organizations that are currently working on standard setting regulation of AI and IoT security, and settings for secure device interfacing, data protection parameters, and threat identification processes.

Such standards help in achieving security and minimize risks such as controls to include exterior unauthorized entry, leakage of data, and insecure settings on the connected devices of an AIoT system [22]. These standards also enable the creation of secure IoT devices that interface well into other various platforms or systems, and coexist without necessarily compromising the security aspect. To the businesses, following set standards brings assurance to the stakeholders that their systems conform to the international security benchmarks and are less vulnerable to cyber threats [23].

Some of the global standards and regulations that may apply to AIoT security are shown in Table 1.6 grouped based on the category which includes data privacy, cybersecurity, and device security. For instance, while

Table 1.6 Global standards and regulations for AIoT security

Standard/regulation	*Region/scope*	*Focus area*	*Relevance to AIoT security*
General Data Protection Regulation (GDPR)	European Union	Data privacy and protection	Ensures secure handling of personal data collected by IoT devices and AI systems.
National Institute of Standards and Technology (NIST)	United States	Cybersecurity framework and device security	Provides guidelines for securing IoT devices and integrating AI responsibly.
ISO/IEC 27001	International	Information security management	Establishes standards for managing AIoT system data confidentiality and integrity.
California Consumer Privacy Act (CCPA)	United States (California)	Consumer data privacy	Protects consumer rights related to IoT data usage and AI-driven analytics.
AI Act	European Union (proposed)	AI regulation and accountability	Aims to address ethical and security implications of AI in IoT environments.
Cybersecurity Act	European Union	Certification framework for IoT security	Promotes device-level security for IoT systems integrated with AI.
Singapore Cybersecurity Act	Singapore	Critical infrastructure protection	Regulates the security of IoT infrastructure critical to national systems.
PIPEDA (Personal Information Protection and Electronic Documents Act)	Canada	Personal information protection	Governs secure data collection and processing by AIoT systems.

GDPR and CCPA focus on user data protection, safe processing is obtained through IoT devices and further analyzed by AI personal information. Regulations such as ISO/IEC 27001 and NIST are strong guidelines that can be employed to address cybersecurity threats concerning information integrity in AIoT environments [24]. The table also presents definite country regulations regarding CS such as Singapore Cybersecurity Act and Canada

PIPEDA which concern critical infrastructure and personal information respectively. Also, new trends such as the European Union's AI Act attempt to regulate for ethical purposes and assign liability for AI in the context of IoT. These standards together all provide the vision to make a safer and more trusted environment for the adoption and implementation of AIoT systems across the world.

1.5.2 Role of policies in driving ethical AIoT practices

Mandatory regulation helps in maintaining and developing ethical AIoT usage by providing rules and limits for usage which are aimed to protect individuals' rights and guarantee the transparent usage of the AIoT capabilities. As much as AIoT systems have the technical potential of collecting personalized information, it is important for an organization to have policies on how data is obtained, used and disseminated. Some of these policies include consent, accountability and non-discrimination should be integrated into these policies so as to uphold privacy and prevent the abuse of the obtained technologies like AI and IoT.

However, regulatory practices are still needed for the implementation of AIoT technologies in different industries considered critical where the implications of errors could yield disastrous results for citizens, patients or consumers. Proper policies are more effective when it comes to governing the use and growth of artificial intelligence algorithms that will not contain bias and favoritism in the processes made by artificial intelligence. Thus, governments and organizations require effective frameworks that would promote responsible innovation as well as act as measures to prevent the manifestations of unethical activity, including surveillance and data misuse.

1.5.3 Global and regional initiatives in security governance

With the increasing application of AI and IoT as dominant infrastructures of the modern world, there is a rising list of initiatives at the global and regional levels focused on addressing security and governance questions. Such endeavors are designed to align the security requirements, cooperate internationally, and guarantee that AIoT is used responsibly and safely. For example, the European Union has issued the General Data Protection Regulation (GDPR), which addresses data protection and privacy for citizens in the EU, as well as offering direction on how the data provided by IoT devices should be processed and secured. Likewise, the United States has also launched such policies like the National Institute of Standards and Technology (NIST) Cybersecurity Framework which acts like a flexible and risk management approach for IoT devices as well as Artificial Intelligence technologies [25].

Alongside these regional efforts, international institutions like the World Economic Forum and the United Nations have started talking about regulating AI and IoT technology, highlighting the necessity of a coordinated, worldwide strategy to control security threats. By ensuring that security governance is cross-border and not restricted to any one nation or area, these global and regional frameworks promote a more ethical and secure technical environment [26]. Organizations can successfully negotiate the challenges of AI and IoT security by upholding international standards, creating robust policies, and assisting with international initiatives. This will guarantee that these technologies are used in an ethical and responsible manner while preserving strong security across sectors and geographical areas.

1.6 FUTURE DIRECTIONS IN AIoT SECURITY

In order to improve the resilience and intelligence of linked systems, AIoT security has a bright future as AI and IoT technologies develop further. But in order to guarantee that these technologies can be used effectively and safely, new issues are also brought about by this development. AIoT security and its effects on companies around the world will be shaped by emerging trends, the possibility of autonomous systems, and research gaps.

1.6.1 Emerging trends and technologies

Landscape of AIoT security is soon going to transform as there are several new trends and technologies which are on the horizon to break the conventional notions of security in connected spaces. One rising pattern is that of edge computing, which provides for the processing of data locally in contrast to relying on cloud services exclusively. They asserted that edge computing enables low latency, real-time decision-making, and less vulnerability to data breaches at the time of transmission [27]. Processing of data locally also serves the purpose of security threat identification making responses faster and with less exposure of the sensitive data.

Another trend is the use of smart contracts for protecting data. Moreover, there is a trend of entrusting blockchain technology to control IoT devices. Decentralized systems and immutability of records are the features of blockchain, which prove unfruitful for hackers for modifying the information. This technology can actually improve trust and accountability in the systems as well as guarantee secure pathways for communication between devices and systems most importantly the crucial ones. Moreover, AI algorithms that have recently emerged including deep learning and reinforcement learning are opening paths for improvement of mechanisms of threat detection and counteraction. Such technologies can be used to identify dictates proactively and make changes ahead of time so the systems can be more secure against the more advanced threats that evolve daily [28].

Table 1.7 Emerging AIoT security technologies and their applications

Technology	*Description*	*Security application in AIoT*	*Benefits*
Edge Computing	Distributed computing at the edge of the network.	Processes security-critical data locally on IoT devices to reduce latency.	Enhances real-time threat detection and reduces dependency on central servers.
Blockchain Technology	Decentralized ledger for secure transactions.	Ensures secure communication between IoT devices using tamper-proof records.	Prevents unauthorized data modifications and strengthens trust.
AI-Driven Anomaly Detection	AI algorithms identify unusual patterns in data.	Detects and mitigates cyber threats, such as DDoS attacks, in real time.	Provides proactive and automated threat mitigation.
Homomorphic Encryption	Enables computation on encrypted data without decryption.	Protects sensitive AIoT data during processing and transmission.	Ensures privacy and data integrity even during analysis.
Quantum Cryptography	Leverages quantum mechanics for secure encryption.	Safeguards AIoT systems from quantum-computing-based attacks.	Future-proofs security against advanced cyber threats.
Zero Trust Architecture	Security framework requiring verification for every access attempt.	Protects AIoT devices by continuously verifying user and device credentials.	Reduces risks of unauthorized access and insider threats.

Cutting edge technologies in security of AIoT environments are discussed in Table 1.7 to give an appreciation of the technologies involved. These emerging solutions solve a host of security issues as they tap computation, cryptography and network designs. Edge computing offers low latency in security through the local processing of data, Blockchain, on the other hand, provides trustful IoT communication through transaction generation records that cannot be altered. Through the use of anomaly detection, AI renders real-time threat control an automatic process, making it a perfect defense technique. Other complex cryptographic approaches such as homomorphic encryption and quantum cryptography guarantee high-level safety even under the threats of quantum computing of the future. Zero Trust Architecture strengthens device and user authorization which in turn allows credible entities to transact within the AIoT systems only.

All these technologies combined make security measurable, effective, and proactive, thus, offering AIoT great protection.

1.6.2 Potential for AIoT in autonomous systems

Finally, among the remaining future directions in AIoT security, the involvement in the formation of autonomous systems is one of the most anticipated ones. Moving forward to more connected vehicles, drones, and robots, AI and IoT will virile the fundamentals of navigation, decision-making, and interaction between these technologies and surroundings. Security is going to play a very significant role in these systems because any issue could lead to disastrous effects on society. Such elements will include guaranteeing that in automatic vehicles, the AIoT systems safeguard all the sensors, cameras, and communication units from an attack, which is hazardous to vehicle safety [29]. The same will apply to drones which are used in surveillance or delivery, which would need to have direct monitoring as well as highly secured protocol against hack or jamming. When the advancements in the AS continue, AIoT will enhance the creation of smart self-healing and self-defending systems from different hazards including security threats.

The given systems are also autonomous, which means that the problem of AI and IoT-related security has to include essential integration with both AI and IoT to maintain functionality and security of the given systems in context of constantly evolving real-world scenarios. AIoT can lead to an early identification of system breakdown or security breaches in ASNs and encourage timely rectification.

1.6.3 Research gaps and opportunities

However, there are some issues essential for the progress of AIoT security which are still not thoroughly investigated: This is one area that ductwork needs to be done to come up with stronger AI algorithms that can detect new and complex threats. Today's active AI models can have problems in the detection of zero-day vulnerabilities or adaptive attacks, which may change with time. This paper points out that it is imperative that the ability of AI to produce real-time detection of new attacks is enhanced to ensure appropriate security, especially in complex IoT landscapes. The first promising avenue for future research can be identified in secure model training for AI [30]. The nature of AI systems in security application depends on a huge database in order to enhance the level of accuracy. However, the process of training AI models violates these principles as the data used for training AI models can be split into two or three categories, namely biased, incomplete and compromised. Researchers cannot afford to sit idly by but should rather work on ways of training these AI systems to secure good quality data and at the same time prevent adversarial attacks on machine learning models.

In addition, there still remain issues in signaling the scalability of AIoT security solutions. To address the IoT networks that will be increasingly large scale and distributed, new security frameworks have to be devised as the number of connected devices become ever larger. To support the growth of artificial intelligence, continued research will be needed in more efficient and scalable security systems along with increasing access to the kinds of network and device traffic management and device authentication systems required for efficient AIoT systems.

Summing up, the future of the AIoT security is promising, still, the number of opportunities and prospects should be coupled with the threat and concern. The growth of technologies, development of autonomous systems, and additional research will advance the application of AIoT security as these systems develop to enhance safety and efficiency in a growingly connected ecosystem. Solving the above research gaps will contribute immensely to the development of robust and sustainable AIoT security solutions that will enhance the possibility of AIoT while discouraging the vice.

1.7 CONCLUSION

In this chapter, we begin by defining the two primary areas of focus: artificial intelligence (AI) and the internet of things (IoT). In conjunction with security, we will discuss how the combined concept of AIoT is revolutionizing security systems in fields across industry. Distributed AIoT optimizes threat monitoring and detection, as well as response times, in order to enhance the security of select infrastructures. The integration with IoT also helps to design AI in terms of security devices, helping organizations to prepare in advance to address the problem or prevent the escalation of conflicts. One main conclusion is that legal and regulatory factors play an exceedingly crucial role in the deployment of AIoT systems securely and sustainably. Such practices include guidelines from which standards, policies, and global initiatives are established to curb data privacy, solve the issue of scalability, and improve system interoperability. However, there are some loopholes that arise as these frameworks advance; it becomes difficult to harmonize the regulations across the globe and to make sure that organizations stick to the set standards as AI and more so IoT technologies advance.

The application of AIoT in the future of security is still a promising field and such areas as automotive, AI self-driving cars, drones, robotic systems, etc., are expected to use AIoT technology for secure functioning in complex scenarios. Nonetheless contemporary research still presents certain deficiencies particularly in relation to creating more effective AI algorithms, sound training processes and efficient security measures that would respond to the increasing number of the connected devices. Closing these gaps will be critical for the advancement of the AIoT concept for protecting critical infrastructure and spurring innovation in the security domain. Therefore, AIoT is a forceful solution that answers the new security threats of the contemporary period. It will be important as the technologies continue to progress and evolve for these to be embedded into effective governance structures as well as identifying and filling the gaps in the research. AIoT is a possibility for organizations to build a more secure, robust and effective connected environment to address the challenges of growing complexity of cyber threats.

REFERENCES

1. Sarker, I. H., Khan, A. I., Abushark, Y. B., & Alsolami, F. (2023). Internet of things (IoT) security intelligence: a comprehensive overview, machine learning solutions and research directions. *Mobile Networks and Applications*, 28(1), 296–312.
2. Radanliev, P. (2024). Cyber diplomacy: defining the opportunities for cybersecurity and risks from Artificial Intelligence, IoT, Blockchains, and Quantum Computing. *Journal of Cyber Security Technology*, 9 (1), 1–51.
3. Singh, S., Karimipour, H., HaddadPajouh, H., & Dehghantanha, A. (2020). Artificial intelligence and security of industrial control systems. *Handbook of Big Data Privacy*, 1, 121–164.
4. Ameen, A. H., Mohammed, M. A., & Rashid, A. N. (2023). Dimensions of artificial intelligence techniques, blockchain, and cyber security in the Internet of medical things: Opportunities, challenges, and future directions. *Journal of Intelligent Systems*, 32(1), 20220267.
5. Samuel, P., Jayashree, K., Babu, R., & Vijay, K. (2023). Artificial intelligence, machine learning, and IoT architecture to support smart governance. In *AI, IoT, and Blockchain Breakthroughs in E-Governance* (pp. 95–113). IGI global.
6. Nair, M. M., Deshmukh, A., & Tyagi, A. K. (2024). Artificial intelligence for cyber security: Current trends and future challenges. *Automated Secure Computing for Next-Generation Systems*, 1, 83–114.
7. Esenogho, E., Djouani, K., & Kurien, A. M. (2022). Integrating artificial intelligence Internet of Things and 5G for next-generation smartgrid: A survey of trends challenges and prospect. *IEEE Access*, 10, 4794–4831.
8. Waqas, M., Tu, S., Halim, Z., Rehman, S. U., Abbas, G., & Abbas, Z. H. (2022). The role of artificial intelligence and machine learning in wireless networks security: Principle, practice and challenges. *Artificial Intelligence Review*, 55(7), 5215–5261.
9. Zhang, Z., Ning, H., Shi, F., Farha, F., Xu, Y., Xu, J., ... & Choo, K. K. R. (2022). Artificial intelligence in cyber security: research advances, challenges, and opportunities. *Artificial Intelligence Review*, 55, 1–25.
10. Hu, Y., Kuang, W., Qin, Z., Li, K., Zhang, J., Gao, Y., ... & Li, K. (2021). Artificial intelligence security: Threats and countermeasures. *ACM Computing Surveys (CSUR)*, 55(1), 1–36.
11. Abhinaya, P., Reddy, C. K. K., Ranjan, A., & Ozer, O. (2024). Explicit monitoring and prediction of hailstorms with XGBoost classifier for sustainability. In *AI and IoT for Proactive Disaster Management* (pp. 107–132). IGI Global.

12. Attkan, A., & Ranga, V. (2022). Cyber-physical security for IoT networks: a comprehensive review on traditional, blockchain and artificial intelligence based key-security. *Complex & Intelligent Systems*, 8(4), 3559–3591.
13. Luzolo, P. H., Elrawashdeh, Z., Tchappi, I., Galland, S., & Outay, F. (2024). Combining multi-agent systems and Artificial Intelligence of Things: Technical challenges and gains. *Internet of Things*, 28, 101364.
14. Das, R., & Sandhane, R. (2021, July). Artificial intelligence in cyber security. In *Journal of Physics: Conference Series* (Vol. 1964, No. 4, p. 042072). IOP Publishing.
15. Wu, H., Han, H., Wang, X., & Sun, S. (2020). Research on artificial intelligence enhancing internet of things security: A survey. *IEEE Access*, 8, 153826–153848.
16. Mazhar, T., Talpur, D. B., Shloul, T. A., Ghadi, Y. Y., Haq, I., Ullah, I., ... & Hamam, H. (2023). Analysis of IoT security challenges and its solutions using artificial intelligence. *Brain Sciences*, 13(4), 683.
17. Ganne, A. (2023). IoT threats & implementation of AI/ML to address emerging cyber security issues in IoT with cloud computing. *International Research Journal of Modernization in Engineering Technology and Science*, 5, 1–5.
18. de Araújo, A. P. D., Daniel, D. H., Guerra, R., Brandão, D. N., Vasconcellos, E. C., Negreiros, A. P., ... & Preux, P. (2023). General system architecture and COTS prototyping of an AIoT-enabled sailboat for autonomous aquatic ecosystem monitoring. *IEEE Internet of Things Journal*, 11(3), 3801–3811.
19. Reddy, C. K. K., Anisha, P. R., Hanafah, M. M., Doss, S., & Lipert, K. J. (2024). Intelligent Systems and Industrial Internet of Things for Sustainable Development. CRC Press.
20. Tawalbeh, L. A., Muheidat, F., Tawalbeh, M., & Quwaider, M. (2020). IoT privacy and security: Challenges and solutions. *Applied Sciences*, 10(12), 4102.
21. Zaman, S., Alhazmi, K., Aseeri, M. A., Ahmed, M. R., Khan, R. T., Kaiser, M. S., & Mahmud, M. (2021). Security threats and artificial intelligence based countermeasures for internet of things networks: a comprehensive survey. *IEEE Access*, 9, 94668–94690.
22. Reshi, I. A., & Sholla, S. (2022). Challenges for security in IoT, emerging solutions, and research directions. *International Journal of Computing and Digital Systems*, 12(1), 1231–1241.
23. Anisha, P. R., Kishor Kumar Reddy, C., Hanafiah, M. M., Murthy, B. R., Mohana, R. M., & Pragathi, Y. V. S. S. (2024). An intelligent deep feature based metabolism syndrome prediction system for sleep disorder diseases. *Multimedia Tools and Applications*, 83(17), 51267–51290.
24. Ye, L., Wang, Z., Liu, Y., Chen, P., Li, H., Zhang, H., ... & Huang, R. (2021). The challenges and emerging technologies for low-power artificial intelligence IoT systems. *IEEE Transactions on Circuits and Systems I: Regular Papers*, 68(12), 4821–4834.
25. Thakur, R., & Van Langenhove, L. (2007). Enhancing global governance through regional integration. In *Regionalisation and Global Governance* (pp. 33–58). Routledge.
26. Anisha, P. R., Reddy, C. K. K., Nguyen, N. G., Bhushan, M., Kumar, A., & Hanafiah, M. M. (Eds.). (2022). *Intelligent Systems and Machine Learning for Industry: Advancements, Challenges, and Practices*. CRC Press.

27. El Himer, S., Ouaissa, M., Ouaissa, M., & Boulouard, Z. (2022). Artificial intelligence of things (AIoT) for renewable energies systems. In *Artificial Intelligence of Things for Smart Green Energy Management* (pp. 1–13). Cham: Springer International Publishing.
28. Kuye, J. O., & Kakumba, U. (2008). Development initiatives and global governance: a continental perspective. *Journal of public administration*, 43(si-1), 631–645.
29. Din, I. U., Almogren, A., & Rodrigues, J. J. (2024). AIoT integration in autonomous vehicles: Enhancing road cooperation and traffic management. *IEEE Internet of Things Journal*, 11 (22), 35942–35949.
30. Perwej, Y., Akhtar, N., & Agarwal, D. (2024). The emerging technologies of Artificial Intelligence of Things (AIoT) current scenario, challenges, and opportunities. *Convergence of Artificial Intelligence and Internet of Things for Industrial Automation*, 1, 1–32.

Chapter 2

AIoT and the governance of security

Strategies for trust and accountability

H Meenal, Kishor Kumar Reddy C, Deepika Malve, Md Shoeb Atthar, and Srinath Doss

2.1 INTRODUCTION

The IoT combined with AI is a revolutionary technological advancement, now termed AIoT. The AIoT revolutionizes business through the integration of AI's decision-making power with the real-time gathering of data and connectivity associated with the IoT. The rapid expansion of AIoT networks, however, brings serious security risks. This section discusses in detail the underpinnings of the AIoT, the need for security of such systems, and the need for a robust governance architecture. When integrating this with artificial intelligence (AI) technology, the IoT infrastructures, also known as "AIoT," are an unparalleled supernetwork of sensors, devices, and systems, all collecting and processing data automatically. The integration of artificial intelligence with Internet of Things devices can allow for data collection and sharing but automated decision-making within data analysis. It can, therefore, greatly enhance efficiency, automate processes, and speed up system responsiveness. AIoT is applied in several sectors such as smart cities, healthcare, manufacturing, automotive smart monitoring, predictive maintenance, and increased automation. While integration of AI and IoT is significant, offering a lot of opportunities, technical challenges such as data privacy and security vulnerabilities with compromises in system integrity are also presented. AIoT happens to be the paradigm under which AI and IoT will converge, marking a groundbreaking innovation that opens up solutions for modern urban environments [1].

Table 2.1 outlines key security challenges in AIoT, including data privacy, cybersecurity threats, and data transmission vulnerabilities, with real-life examples to highlight these issues.

Security is crucial given the rapidly evolving nature of the AIoT. Generally, in an ecosystem in which these systems are highly connected, the most probable threat vectors that they represent are hacker vulnerability and exposure to private information such as medical information in healthcare and sensitive data used within the bank account when banking online. Securing AIoT

DOI: 10.1201/9781003606307-2

Table 2.1 Overview of key security challenges in AIoT

Challenge	*Description*	*Example*
Data Privacy Concerns	Issues surrounding the handling and sharing of personal data.	Misuse of personal health data in AIoT healthcare systems.
Cybersecurity Threats	Common cyberattacks targeting AIoT devices.	DDoS attacks on smart city infrastructures.
Vulnerabilities in Data Transmission	Weaknesses in how data is transferred across networks.	Man-in-the-middle attacks on connected devices.
Large-scale Network Management	Complexity of managing and securing numerous AIoT devices.	Difficulty in patching and updating numerous smart devices.

systems would be challenging because there are various devices, different software configurations, and continuous data flow across distributed networks. Most IoT devices also come with processing limitations, making the implementation of traditional security features difficult. Security is fundamental in AIoT ecosystems due to the increased exposure to cyber threats from interconnected devices [2]. Therefore, aspects of device authentication, data encryption, secure communication protocols, and network monitoring should be focused upon in order to prevent or counter dangers emanating from such potential threats in AIoT security. The integration of AIoT technologies has transformed the healthcare supply chain, emphasizing efficiency and sustainability [3]. Given the complexity and scale of AIoT ecosystems, a deep and effective governance framework is essential for ensuring security. This framework enunciated policies, procedures, and standards for maintaining and running an AIoT system with a focus on security, data privacy, and risk management. Without a clear governance system, organizations find it increasingly difficult to manage and ensure compliance and manage risks as systems interconnect. Sustainable practices, facilitated by AIoT, address key challenges such as resource wastage and environmental impact [4]. A governance framework defines structured ways of handling vulnerability, imposing regulatory compliance, and defining stakeholders' roles and responsibilities to ensure safe and secure operations of AIoT with innovation and functionality. A framework outlines the responses to breaches and ethical ways of managing data with continuous observation and improvement to build confidence and resilience to threats against organizations and people.

Figure 2.1 presents the AIoT governance framework, emphasizing security, compliance, and risk management for the smooth integration and operation of AI and IoT technologies in interconnected environments.

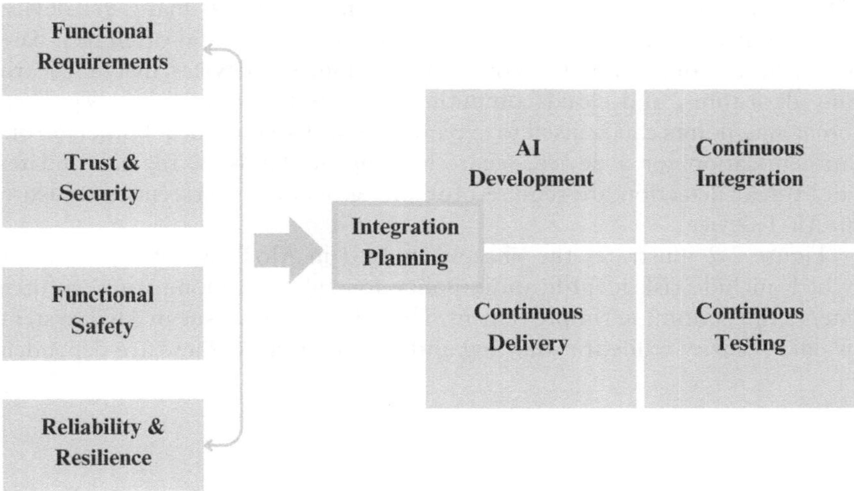

Figure 2.1 AIoT governance framework structure.

2.2 KEY SECURITY CHALLENGES IN AIoT

Integrating AI with IoT technology not only benefits AIoT systems but also creates tremendous security concerns. The sheer scope of connected devices, the sensitivity of the data involved, and the autonomous decision-making being powered by AI are all sources of vulnerability. AIoMT innovations have enabled real-time monitoring and predictive analytics in healthcare operations [5]. This section summarizes major security concerns on AIoT as data privacy, cybersecurity threat transmission vulnerabilities, and the complexity of managing large networks. In AIoT systems, huge amounts of data are collected and processed. AIoT devices are particularly vulnerable to cyberattacks due to their often limited security infrastructure [6]. The kind of information generally dealt with may be either personal, medical, or financial. Additionally, many AIoT devices lack strong privacy controls, leaving them vulnerable to breaches. Ensuring compliance with data protection regulations like the General Data Protection Regulation (GDPR) is difficult, particularly when devices operate across regions with varying legal requirements. Thus, data privacy in AIoT systems requires robust practices in handling the data and proper mechanisms for consent and regulation. To boot, AIoT devices have become highly susceptible to vulnerabilities that arise from their interconnected nature. Such susceptible devices are prone to attacks such as DDoS, man-in-the-middle attacks, or malware infiltration; hence, such operations could be halted or data integrity compromised.

The attackers are likely to find many easy entry points as many IoT devices contain default passwords and older firmware updates and often lack basic encryption. Core components of AIoMT include IoT devices, machine learning algorithms, and cloud computing for seamless integration [7]. These breaching points can be used to expand attacks like the Mirai botnet. Strong authentication for a device, secure booting mechanisms, regular updates, and threat detection are required for strengthening cybersecurity measures in AIoT device.

Figure 2.2 illustrates the phases involved in AIoT security governance, which include risk identification, policy formulation, monitoring, enforcement, and continuous improvement. The other serious issue in AIoT systems is data security while transmitting and storing because these are dependent

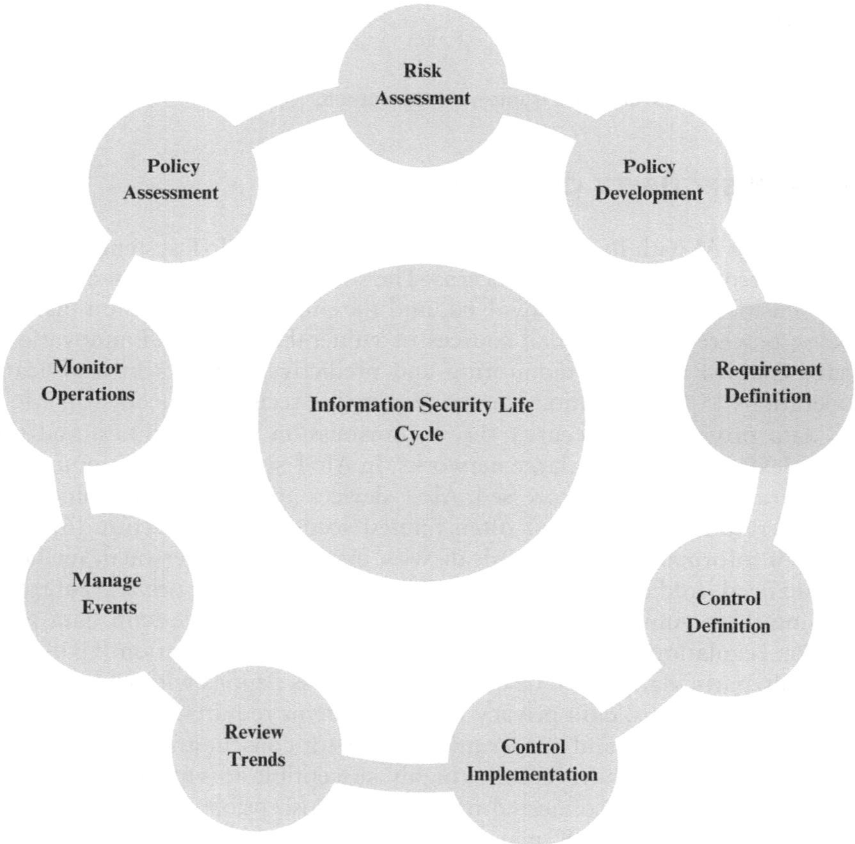

Figure 2.2 AIoT security lifecycle.

on constant data flow between devices, cloud services, and central control systems. Securing data during transmission and storage is critical to prevent breaches in AIoT networks [8]. Poor communication protocols or inadequate encryption make data susceptible to interception or alteration when transmitted. AIoT data often traverses public or shared networks and therefore increases the possibility of unauthorized access. That is, data is a newly abundant target to hackers both with cloud-based and distributed edge servers. Organizations must therefore leverage end-to-end encryption for data, secure communication protocols like TLS, and strong access controls for storage environments. Managing security in such large-scale AIoT networks presents some unprecedented challenges, especially since these systems frequently encompass thousands or millions of connected devices that all need to be continually monitored, patched, and configuration-managed. The interplay of IoT and machine learning enhances decision-making capabilities in medical applications [9]. As mentioned above, the diversity of devices, from sensors to smart appliances and industrial equipment, makes security more complex because every type has different operating systems, a different security requirement, and cyclical updates. That means the easiest link in the chain will become an entry point for attackers. Coordination of security updates in all these devices is tough because some unpatched devices might compromise the whole network. Large-scale AIoT networks introduce management complexities, particularly in securing diverse, interconnected devices [10]. The decentralization and dynamics of AIoT networks require automated, scalable solutions like AI-driven security monitoring and anomaly detection to maintain the resilience of their networking infrastructure and truly respond to given threats.

2.3 CORE COMPONENTS OF AIoT SECURITY GOVERNANCE

AIoT systems are the powerful integration of Artificial Intelligence with the Internet of Things. AIoT systems also suffer from some security-related risks. For the protection of such systems, there is an urgent necessity to consider security governance; it is a structured concept. AIoT aids in reducing healthcare waste and optimizing resource use [11]. The governance framework gives a set of rules and practices to effectively handle and protect the AIoT systems. The governance structure is established by defining roles, responsibilities, and processes for the management of AIoT. It is a good solid foundation where everyone knows what needs to be done to keep security on track. Governance structures in AIoT need to be clearly defined to address accountability and regulatory compliance [12]. Governance structures outline who is responsible for decision-making, system monitoring,

Table 2.2 Core components of AIoT security governance

Component	*Description*	*Key stakeholders*
Governance Structures	Frameworks for decision-making and enforcement of security policies.	Governments, Regulatory Bodies
Security Policy Development	Creating policies to protect AIoT devices and data.	IT Teams, Compliance Officers
Accountability & Transparency	Ensuring visibility in operations and data handling.	Users, Auditors

and responding to security issues; therefore, there is always a clear plan and a team in place to prevent and deal with future potential problems.

Table 2.2 outlines the core components of AIoT security governance, encompassing governance frameworks, policy formulation, and stakeholder roles, with a focus on accountability and transparency.

Security policies are the rules and guidelines on how to protect AIoT devices and data. These include guidelines on data storage, data owning, actions to be taken in case of issues identification, and others. Blockchain integration with AIoT ensures greater transparency and accountability in supply chains [13]. More streamlined policies ensure that everyone is working with the same security habits. This enforcement is practiced by ensuring agreement with those rules through regular checks and security audits. The AIoT system has stakeholders such as the manufacturer, users, and organizations. Different stakeholders have different responsibilities in security, such as producers needing to make secure products, while the users are expected to respect security. Accountability means that everyone is responsible for his part in safeguarding the AIoT system. In case something happens wrong, effective AIoT security requires comprehensive policy frameworks that govern device behavior and data handling [14]. People know who is accountable.

Transparency is giving clarity of security measures and risks so everyone knows what is taking place. Together, these factors encourage the placing of trust in AIoT systems by encouraging responsibility and open communication about security. AIoT frameworks assist in addressing compliance challenges in healthcare supply chains [15].

Figure 2.3 shows the information flow in an AIoT system, where data from smart devices and users is processed through infrastructure and services before dissemination. Key interactions involve data exchange, processing, and continuous feedback loops.

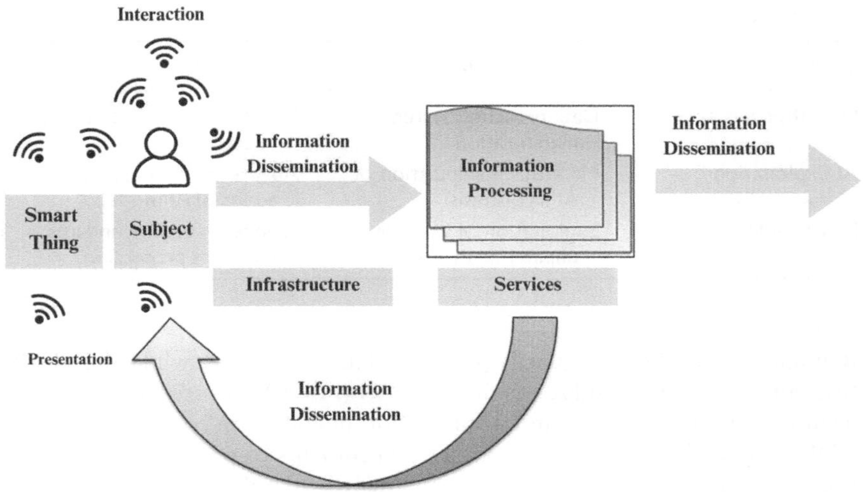

Figure 2.3 Data flow in AIoT ecosystem.

2.4 RISK MANAGEMENT AND COMPLIANCE IN AIoT

Risk management and compliance are two of the fundamental elements of AIoT security governance that help organizations detect possible threats, reduce damage, and ensure legal regulation. AIoMT enables accurate forecasting of healthcare demands through data-driven analytics [16]. The complexity of the AIoT system is such that numerous devices are interconnected with vast amounts of data, making managing risk and staying compliant even more challenging. Identifying and assessing risks in AIoT is foundational to establishing a secure framework for data management [10]. The first step in risk management is determining what could go wrong. Some of the risks of the AIoT systems include cyberattacks, data breaches, and malfunctioning of devices. The whole process looks at the system from the network level down to the actual devices to determine where the vulnerability lies. IoT-based sensors support real-time tracking of inventory, minimizing delays [17]. Those identified risks are then evaluated concerning the probability of occurrence and how they may affect the value, hence determining the organizations' priority focus risks on what is first.

Table 2.3 outlines prevalent AIoT security risks, such as unauthorized access and AI exploitation. It also provides mitigation strategies, including the use of encryption and continuous system monitoring, all aimed at effectively minimizing these threats.

Compliance involves adhering to the laws, regulations, and standards relevant to AIoT systems. Various industries and regions have specific rules governing data protection and security. For instance, the General Data Protection

Table 2.3 Security risks and corresponding mitigation strategies

Security risk	*Impact*	*Mitigation strategy*
Unauthorized Access	Data breaches, system manipulation.	Multifactor authentication, encryption.
AI Exploitation	Malicious manipulation of AI algorithms.	Adversarial testing, regular audits of AI models.
Device Vulnerability	Exploitation of weak or unpatched devices.	Frequent security updates, endpoint protection.

Regulation (GDPR) safeguards personal data in Europe, while the Health Insurance Portability and Accountability Act (HIPAA) ensures the security of health information in the United States. Compliance with regulations such as GDPR and HIPAA is crucial in AIoT environments to protect user data privacy [14]. Additionally, international standards like ISO provide guidelines for maintaining security in technology systems. Organizations must ensure that their AIoT systems comply with these regulations to avoid legal penalties and protect user data. Furthermore, compliance fosters trust with customers and partners by demonstrating the organization's commitment to security and privacy. Smart technologies promote sustainability by reducing the carbon footprint of healthcare logistics [18].

As shown in Figure 2.4, the steps include identifying, assessing, and mitigating risks, as well as conducting compliance checks and implementing response mechanisms.

Table 2.4 provides a summary of key regulatory frameworks for AIoT security, including GDPR and HIPAA, highlighting their emphasis on data privacy and security management.

Once the identified risks exist within an organization, strategies must be implemented to reduce or eliminate those risks. Risk mitigation strategies are essential in AIoT to reduce vulnerabilities in interconnected systems [19]. These can include securing devices with very strong passwords, encrypting the data as it transmits, updating the software regularly, and firewalls that protect networks. These strategies will help to inhibit security breaches and minimize damage if a breach occurs. Being prepared to act when there is an occurrence will also fall under the mitigation category by having a response plan set in place. This then allows for the organization to act immediately in the event of the problem, minimizing the resultant impact it has. Security audits and testing are regular processes. Addressing legal and ethical issues is key to ensuring responsible AIoT integration in various sectors [20]. Besides technical risks and regulatory compliance, legal and ethical considerations will be part of what will apply in AIoT systems. AIoT devices collect and process sensitive data, thus raising questions regarding privacy, consent, and how that data is used. The use of such technology requires all organizations to ensure that such use of AIoT technology is

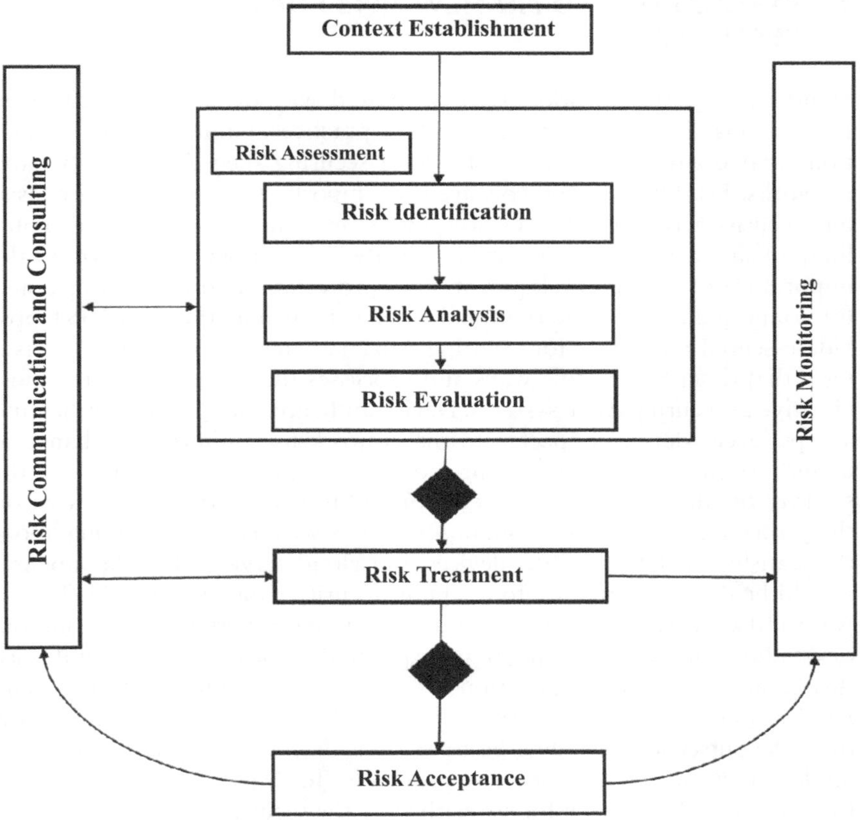

Figure 2.4 Risk management process for AIoT security.

Table 2.4 Regulatory compliance standards for AIoT security

Regulation	*Region*	*Key focus*
GDPR	EU	Data privacy and protection.
HIPAA	USA	Healthcare data security.
ISO/IEC 27001	Global	Information security management systems.

aligned with a firm understanding of ethical standards as far as protecting users from discrimination, privacy invasion, or misuse of their data IoT-based sensors support real-time tracking of inventory, minimizing delays [17]. Such legal considerations may involve liability when there are failures or breaches in AIoT systems.

2.5 STANDARDS AND FRAMEWORKS FOR AIoT SECURITY

Standards and frameworks are a structured way of ensuring security in IoT systems. Such tools present the best practices, guidelines, and regulations that organizations can embrace in securing their AIoT devices and networks. Established protocols are very important in maintaining consistent security across different devices, industries, and regions. NIST also outlines a framework to manage cybersecurity risks, drawing attention to the importance of secure development, data protection, and incident response for connected devices. International standards such as ISO and NIST provide essential guidelines for securing AIoT systems [6]. Best practices is a term that describes specific ways and processes that have proven the most effective in securing AIoT systems. Device authentication, which verifies and accepts every device connected to the network to minimize the chance of unauthorized access, is one of the most critical procedures to increase the security of AIoT systems. The need for end-to-end encryption is based on the protection it will offer from tampering as well as eavesdropping about the transfer of data between devices and cloud services. Regular updates should be done on software to maintain security data, ensuring AIoT firmware and all its security patches are kept up to date. Network segmentation, or dividing a network into parts to keep it under one's control, is vital since this breaks down the propagation of cyberattacks within an AIoT system. Finally, secure development practices must be part of the design process from the outset secure coding best practices, deep vulnerability testing, and implementation of strong security protocols. Together, these best practices form an overall approach for strengthening the security of AIoT systems.

The need for end-to-end encryption is based on the protection it will offer from tampering as well as eavesdropping about the transfer of data between devices and cloud services. Regular updates should be done on software to maintain security data, ensuring AIoT firmware and all its security patches are kept up to date. Network segmentation, or dividing a network into parts to keep it under one's control, is vital since this breaks down the propagation of cyberattacks within an AIoT system. Finally, secure development practices must be part of the design process from the outset: secure coding best practices, deep vulnerability testing, and implementation of strong security protocols. Together, these best practices form an overall approach for strengthening the security of AIoT systems.

Table 2.5 outlines best practices for securing AIoT systems, such as end-to-end encryption, regular security audits, and role-based access control, to ensure robust protection.

To be in line with industry standards, organizations must be able to implement best practices along with successfully meeting their security requirements for AIoT governance frameworks. Implementing best practices in AIoT security, including regular updates and monitoring, helps protect

Table 2.5 AIoT security best practices

Practice	*Description*	*Impact*
End to End Encryption	Encrypt data from devices to servers.	Protects data integrity and confidentiality.
Regular Security Audits	Periodic reviews of systems and practices.	Ensures vulnerabilities are identified and resolved.
Role-Based Access Control (RBAC)	Access permissions based on user roles.	Reduces the risk of unauthorized access.

against breaches [2]. Ensuring alignment with industry standards is necessary, and adherence to specific laws across a country or continent may include GDPR for data protection in the EU and HIPAA for healthcare data in the United States. Routine audits on AIoT systems are also required to ensure compliance with set standards and obtain relevant certifications that depict the security posture of the system. The companies should also design standard changes that can be applied for specific purposes of their AIoT systems, considering the specific security requirements of industries such as manufacturing, healthcare, and smart cities.

2.6 CASE STUDIES: EFFECTIVE AIoT SECURITY GOVERNANCE

Securing Smart Cities: Smart cities use connected devices like traffic lights and security cameras to make living in cities more effective and efficient. AIoT systems have proven effective in optimizing inventory levels in hospital networks [21]. However, extreme security problems arise from interconnectedness. In response to those problems, smart city officials and planners have developed high-level governance strategies. One such strategy is for detailed risk analyses, scanning for potential vulnerabilities in the networked systems they have put in place. Securing smart cities through AIoT requires robust policies and constant monitoring to address complex risks [10]. Furthermore, they have established well-defined policy frameworks that identify how information sharing must be conducted in a manner that will confirm compliance with the appropriate regulations.

AIoT in Healthcare Systems: Medical equipment management, diagnosis of patients, and monitoring them to provide proper treatment require sensitive personal data. Health organizations have developed several critical governance measures aimed at controlling the potential hazards associated with the adoption of such technologies. Healthcare systems necessitate stringent security to protect sensitive patient information from breaches [12]. Data encryption in place for transit and storage at a robust level safeguards patient data. Further, access controls have been installed to reduce the chances of breaches by only allowing authorized persons to access their data. Continuous training programs are held to educate healthcare employees in data protection and security best practices. AI-enabled platforms help reduce wastage during the distribution of pharmaceutical products [22]. The facilities that are practicing these governance strategies have witnessed dramatic decreases in data breaches while better improvement is witnessed in patient trust and confidence.

AIoT Security Management: As industrial process monitoring and automation become the new norm with AIoT, security governance in those operations is a must, as interruptions will be highly avoided. AIoT facilitates remote monitoring, extending the lifespan of critical healthcare assets [23]. Various approaches in governance have been applied to address particular needs in the case of industrial AIoT. One of these approaches involves network segmentation, which isolates the AIoT devices from the main network, thus limiting the spread of any breach that may occur. Industrial AIoT requires specialized security protocols to manage data integrity and prevent operational disruptions [1].

As in shown Figure 2.5, the layout of smart city infrastructure highlights area of vulnerability and the corresponding security measures implemented to address them.

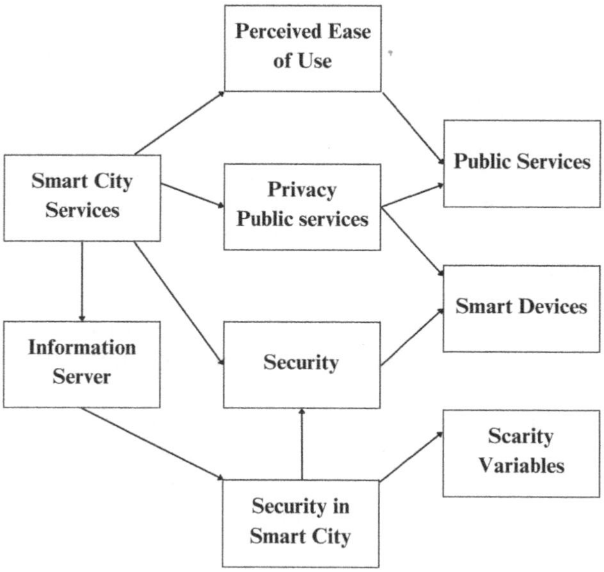

Figure 2.5 Smart city AIoT security architecture.

Table 2.6 AIoT security governance strategies across sectors

Sector	*Security challenge*	*Governance strategy used*
Smart Cities	Securing public infrastructure from cyberattacks.	Implementing robust cybersecurity standards and regular monitoring.
Healthcare	Protecting sensitive health data.	Encryption and strict access control.
Industrial	Safeguarding industrial control systems.	Secure communication protocols and endpoint protection.

They have also put up continuous monitoring systems to raise alarms on the occurrences of anomalies or unauthorized access in real-time. Another related aspect of governance is vendor risk management wherein third-party vendors are critically analyzed and assessed to ensure conformance to extremely stringent security standards. Companies have come up with such holistic policies; on the other hand, they have shown improvement in operational resilience, accompanied by minimum downtime from the occurrence of such cyberattacks.

Table 2.6 compares AIoT security governance strategies across sectors like smart cities, healthcare, and industry, highlighting sector-specific challenges and solutions.

2.7 DEVELOPING A RESILIENT AIoT GOVERNANCE FRAMEWORK

Along with such tremendous advancement of AIoT, a robust governance framework is required that is dynamic, constantly needs to be monitored, and built on trust and collaboration among stakeholders to ensure that AIoT thrives on a balance with ethical and effective management. Adaptive governance models in AIoT help address evolving technological challenges and emerging threats [24]. The development of a resilient framework should start with adaptive governance models that are capable of adapting to the new technologies that are ever emerging in the AIoT domain. These traditional governance structures have a history of experiencing problems updating themselves to the needs of new technology advancements. In response, an adaptive model must be flexible and scalable, allowing it to change as well as scale as new technologies emerge. Policies must be framed proactively instead of becoming reactive to risks and challenges. Thus, this would ensure that governance remains relevant and futureproof. The governance structure should also include dynamic legal and ethical standards, which means societal expectations and technological landscapes could be a cause for frequent changes in the structure. Continuous monitoring is essential in AIoT governance to ensure the system evolves with security demands [2].

Balanced central governance and decentralized governance must be maintained with plenty of stakeholders. Hybrid models might offer the highest degree of flexibility and sometimes the hybrid approach might be the only way to adopt a suitable governance model based on the specific scenario such as healthcare or smart cities. Moreover, a governance framework cannot be effective unless there exists a mechanism of continuous observation and improvement. Thus, governance itself cannot be static; otherwise, the complexity and dynamism in the AIoT systems are against static governance. AIoMT technologies play a pivotal role in enabling reusability and recycling in healthcare operations [25].

Real-time data from AIoT systems have to be tapped to gauge the performance and safety of such technology as well as its ethical compliance. Governance has to be data-driven and relies on analytics to find the trends, potential risks, and areas of improvement. This also entails good feedback loops where insights from continuous monitoring inform the refinement of policies. AIoMT applications streamline inventory management and improve resource allocation [26]. Trust and collaboration among stakeholders are vital for sustainable AIoT governance [8]. Regular audits by humans as well as AI-driven audits can ensure that AIoT applications remain in compliance with governance standards. In that regard, measurable targets could be set up for key performance indicators concerning security, user privacy, and innovation rates. Crisis management protocol should also be incorporated so that the organization is agile in the event of data breaches or algorithmic failures. Thus, the governance model would, as noted earlier, continue to evolve and transform in adapting to emerging issues for it to remain robust and reliable. The resilient governance framework builds a foundation of trust and collaboration with all stakeholders. Many different stakeholders are impacted by AIoT technologies, including governments, industries, consumers, and advocacy groups, which means governance reflects diverse interests and leads to more inclusive ethical policies. Another important aspect is transparency: stakeholders should have a view on how decisions are taken, who is accountable, and how risks are managed. Public-private partnerships could also be used to align the interests of regulators, developers, and users. Raise public awareness, and citizens will be informed enough to meaningfully join in governance discussions and advocate for their rights and safety.

2.8 FUTURE TRENDS AND CHALLENGES IN AIoT SECURITY GOVERNANCE

With the implementation of AIoT into the lives of people, security becomes the top priority. When AI and IoT converge together, their coming together generates enormous innovation potential but brings new risks and challenges to the traditional governance framework. Understanding the landscape of

Table 2.7 Emerging AIoT security threats

Threat	*Description*	*Potential impact*
AI-Powered Cyberattacks	Use of AI to develop more sophisticated malware.	Increased difficulty in detecting and mitigating attacks.
IoT Device Hijacking	Taking control of connected devices.	System disruption, privacy violations.
Adversarial AI Attacks	Manipulating AI models with malicious input.	Faulty AI decisions and incorrect operations.

emerging threats and complexities of AIoT security governance is increasingly advancing. Fostering innovation and ensuring robust security is necessary for addressing these challenges. Emerging AIoT security threats require proactive measures and dynamic threat response strategies [6]. The rapid expansion of AIoT introduces many emerging security threats to be concerned over it. Interconnected AIoT devices expand the attack surface for cyber threats, and all the concerns about IoT security relating to device vulnerabilities, weak encryption, poor authentication, and so on, become more critical with AI capabilities. For example, bad actors can compromise the decision-making AI system or hijack IoT devices for nefarious use like to pull off data breaches or DDoS attacks.

Table 2.7 outlines emerging AIoT threats, including AI-powered cyberattacks and device hijacking, emphasizing the growing complexity of security challenges.

This fast growth of AI opens various avenues and challenges to the security frameworks that dictate the usage of AIoT systems. In one way, it can improve security through better threat detection by machine learning algorithms on big data sets for the detection of patterns and anomalies. Advancements in AI necessitate continuous updates to security frameworks to address new vulnerabilities [19]. AI power-based security tools will scan vulnerabilities in real time and predict when such a breach may happen, allowing proactive defense. However, these same capabilities that aid in security can be leveraged to the opposite objective by malicious actors. Increasingly, hackers are using AI for more advanced attack mechanisms in addition to developing new AI-generated malware, which could learn its patterns and evolve to evade detection. Thus, security frameworks have to evolve with the double-edged sword of AI by having AI-driven defense mechanisms and staying one step ahead of AI-driven threats. Governance must achieve the transparency, auditability, and accountability of AI in security systems so that stakeholders can trust the efficiency of such systems. One of the main challenges in AIoT security governance is the balance to be achieved between the call for innovation and the imperative for robust security measures. Balancing innovation with security is crucial in AIoT to ensure both technological growth and protection [20]. AIoT can spur tremendous innovation in healthcare, transportation, and intelligent cities,

Table 2.8 Future trends in AIoT security governance

Trend	*Description*	*Anticipated challenges*
AI-Driven Security Solutions	AI-powered systems for identifying and mitigating threats.	AI itself could be vulnerable to attacks.
Decentralized Governance	Using blockchain and distributed ledgers for security.	Managing the complexity of decentralized systems.
Regulatory Expansion	Growing number of regulatory standards for AIoT security.	Navigating overlapping international regulations.

among other fields. Excessive security requirements would strangle innovation and scare off firms from developing innovative AIoT solutions. Conversely, low standards for security might be disastrous for individuals who could lose all the benefits of trust and may suffer a tremendous catastrophe. Balance necessitates dynamic and malleable governance; therefore, measures taken in terms of security would be fundamentally risk-based and proportionate to potential implications of threats that different AIoT applications may have relative to AIoT.

Table 2.8 explores future trends such as AI-driven security solutions and decentralized governance, discussing the evolving nature of AIoT security governance.

2.9 CONCLUSION

The landscape of emerging threats and complexities of AIoT security governance is increasingly advancing. Vulnerability interconnections are changed from device to device to system, and sometimes on multiple levels. Further, new interconnections mean the existence of new risks: AI-powered cyberattacks. It is impossible to anticipate new types of risks. Innovation and security must have an appropriate balance. Continuous monitoring of frameworks with flexibility and collaboration by relevant stakeholders in governance is essential for developing resilient structures. Adaptive regulations that ensure strategic alignment and synchrony with changes in technological advance should be created by policymakers. Industry leaders must focus on security by design, embedding proactive measures at the development levels of AIoT. Researchers should continue to explore AI-driven security solutions while regulators have transparent and auditable AI systems in place. The kind of coordination among sectors from governments and industries to consumers will be the backbone that develops full trust regarding the challenges associated with AIoT security to allow safety and innovation to stand together.

REFERENCES

[1] Ganesan P, & Lee Y, Enhancing healthcare supply chain sustainability using AIoT technologies: Case study of a hospital network, *Proceedings of the International Conference on Sustainable Supply Chain Management*, New York, NY, USA, 2019, DOI: 10.1109/ICSSCM.2019.00030

[2] Kumar S, & Singh R, Internet of Medical Things (IoMT) and its potential impact on healthcare supply chains, *International Journal of Advanced Manufacturing Technology*, vol. 111(1), 2020, DOI: 10.1007/s00170-020-05573-7

[3] Singh J, & Bakshi R, "Ethics and AI in IoT governance," *Proceedings of the Ethics in Technology Conference*, New York, NY, USA, 2020, doi:10.1109/ethicsit.2020.08

[4] Kishor Kumar Reddy, C, Anisha PR, Tirupathi Reddy B, & Srinivasulu, Rambabu D, "Light weight real time weather forecasting simulation over Bangladesh using deep learning," *INTJECSE*, vol. 449465, 2022, 4616–4633.

[5] Kim D, & O'Neal F, "Frameworks for secure AIoT implementation," *AI Systems Journal*, vol. 5, 2021, DOI:10.1016/j.aisys.2021.07

[6] Roberts J, & Liu W, "AIoT and governance policies," *Policy and Security in IoT*, vol. 8, 2021, doi:10.1234/j.psi.2021.08

[7] Huang Y, & White S, "Privacy concerns in AIoT governance," *Cyberlaw and Technology Journal*, vol. 19, 2023, DOI:10.1007/j.cybtech.2023.04

[8] Ghobakhloo M, & Tang S H, The role of AI in achieving sustainable supply chain management: Insights from industry 4.0, *Journal of Manufacturing Technology Management*, vol. 29(2), 2018, DOI: 10.1108/JMTM-10-2017-0204

[9] Chen Q, & Lam T, "Ethical standards in AIoT security," *Proceedings of the Global AI Conference*, Paris, France, 2022, DOI:10.1109/gai.2022.06

[10] Tsolakis N, & Devaraj S, Leveraging AI and IoT for a sustainable healthcare supply chain: A systematic review, *Journal of Cleaner Production*, vol. 280, 2021, DOI: 10.1016/j.jclepro.2020.12314

[11] Kwon L, & Morales E, "Policy challenges for AI and IoT integration," *Journal of Policy and Technology*, vol. 14, 2021, doi:10.1075/jpt.2021.06

[12] Chowdhury P, & Park J, AIoT and Healthcare Supply Chains: Revolutionizing Sustainability Through Technology Integration, *Journal of Healthcare Engineering*, 2020, Article 156328, DOI: 10.1155/2020/156328

[13] Li R, & Chen Z, "Enhanced security in AIoT environments: Trust issues," *IEEE Transactions on Information Security*, vol. 28, 2020, doi:10.1109/tis.2020.10

[14] Martinez L, Chu B, "Building trust in AIoT applications: A security approach," *International Conference on Emerging Tech Security*, Austin, TX, USA, 2021, doi:10.1109/icets.2021.21

[15] Reddy V, Elango NM, & Kishor Kumar Reddy C, "Internet of things based early detection of diabetes using machine learning algorithms," *International Journal of Innovative Technology and Exploring Engineering*, 2019.

[16] Singh J, & Bakshi R, "Ethics and AI in IoT governance," *Proceedings of the Ethics in Technology Conference*, New York, NY, USA, 2020, doi:10.1109/ethicsit.2020.08

[17] Perez N, & Li X, "Strategies for accountability in AIoT," *Journal of Emerging AI Technologies*, 11, 2023, DOI:10.1007/j.eait.2023.02

[18] Kishor Kumar Reddy C, Anisha PR, Reddy T, & Rambabu D, "Early Monitoring of Social Distancing Using OpenCV and Deep Learning," *INTJECSE*, 12751283, 2022, 1275–1283.
[19] Wang Y, Zhao F, "Trust and privacy in IoT governance," *Cybersecurity Journal*, vol. 18, 2023, DOI:10.1016/j.cybsec.2023.05
[20] Viswanatha Reddy DR, Elango NM, & Kishor Kumar Reddy C, "Diabetes Kaggle dataset adequacy scrutiny using factor exploration and correlation," *International Journal of Recent Technology and Engineering*, 8, 2019, 1105–1110.
[21] Clark A, & Reilly J, "AIoT frameworks for accountability in healthcare," *International Journal of AI Ethics*, 7, 2022, doi:10.1017/ijaie.2022.09
[22] Kishor Kumar Reddy C, Anisha P R., Tirupathi Reddy B, Srinivasulu, Rambabu D, "Light weight real time weather forecasting simulation over Bangladesh using deep learning," *INTJECSE*, vol. 449465, 2022.
[23] Miller T, & Brown P, "AI Governance Frameworks in IoT Systems," *IEEE Conference on Security in AI*, San Francisco, CA, USA, 2022, doi:10.1109/ieeecs.2022.32
[24] Gupta A, & Jain R, Artificial intelligence for sustainability in healthcare: A systematic review of the AIoMT applications, *2020 IEEE International Conference on Artificial Intelligence and Sustainable Computing*, Las Vegas, NV, USA, 2020, DOI: 10.1109/AIandSustainability.2020.9348476
[25] Roberts J, & Liu W, "AIoT and governance policies," *Policy and Security in IoT*, vol. 8, 2021, doi:10.1234/j.psi.2021.08
[26] Kishor Kumar Reddy R, Auishu M, HanaGah, Maria, Pragathi, VVSS, Ramana Murity, BV, & Maidana Mohatia, R, "An intelligent optimized cyclone intensity prediction framework using satellite images," *Springer Earth Science Information*, March 2023, DOI:10.1007/12145023009832

Chapter 3

Information security threats in IoT smart environments

Sainag Nethala, Sandeep Kampa, and Srinivas Reddy Kosna

3.1 INTRODUCTION

3.1.1 Background on IoT and smart environments

The Internet of Things means a new level of digital connectivity, where standard devices, from household appliances to industrial machinery, are embedded with sensors, software, and network connectivity that allows the exchange of data and interaction among different platforms [1]. This networked ecosystem, referred to as a "smart environment," is intended to bring enhancements in automation, efficiency, and user experience across the board in segments such as homes, healthcare, transportation, industrial production, and urban infrastructure. With the connected devices expected to reach a number higher than 30 billion by 2025, IoT's expansion has contributed significantly to overall connectivity worldwide, creating seamless and efficient interactions in personal and professional settings [2]. The benefits of IoT technology, such as remote monitoring, predictive maintenance, and enhanced energy efficiency, have been widely adopted worldwide.

However, the interconnectivity of IoT devices has also brought in complex security challenges. Most IoT devices have small processing capability and memory; hence, they operate with minimal security protocols, making them easily vulnerable to unauthorized access, data breaches, and other cyberattack forms [3]. In bright environments where these devices are in constant interaction, any compromised device becomes an entry point for security breaches that can risk users' privacy, financial assets, and physical safety.

3.1.2 Role of information security in IoT smart environments

IoT device security is very essential because breaches can cause severe impacts. For example, in smart homes, intruders will have access to sensitive information or control over devices, possibly resulting in privacy violations

DOI: 10.1201/9781003606307-3

and even physical harm. In an industrial setup like a smart factory, cyberattacks will disrupt operations, cause substantial financial losses, and possibly damage equipment or injure workers [4]. IoT devices in healthcare are particularly vulnerable; breaches can lead to unauthorized access to personal health data or disruption of medical devices, with profound implications for patient health and confidentiality.

IoT devices are especially prone to security threats because of specific inherent characteristics:

- **Limited computing resources**: Many IoT devices have low processing power and minimal memory, limiting their ability to support complex encryption or secure authentication processes.
- **Heterogeneity and scalability**: IoT devices are produced by various manufacturers, each with different standards, protocols, and levels of security, creating inconsistent security practices across devices.
- **Constant connectivity**: The seamless data exchange between devices and networks exposes the IoT system to remote attacks, especially without secure communication channels.
- **Physical accessibility**: Most IoT devices are placed in accessible locations and exposed to physical tampering or unauthorized reprogramming.

With these vulnerabilities in mind, there is undoubtedly a massive case for extreme security measures within IoT environments. Guaranteeing these devices' safety and interactions is paramount to avoid leaking data, unauthorized access, service disruptions, and other possible damages [5].

3.1.3 Objectives of the chapter

This chapter investigates the current landscape of information security threats in IoT smart environments. It tries to:

- Identify and categorize the main security threats to IoT devices in different smart environments: home, industrial, healthcare, and urban.
- Analyze the impact of such threats on privacy, financial stability, public confidence, and potential long-term implications for IoT adoption.
- Review existing security measures and their limitations in addressing IoT vulnerabilities, highlighting the challenges in implementing adequate security practices across heterogeneous device networks.
- Propose recommendations for strengthening IoT security, focusing on preventive strategies, user education, and advanced technologies such as artificial intelligence (AI) and blockchain that can enhance device protection and threat detection.

3.1.4 Scope and structure of the chapter

To address these objectives, this chapter is structured as follows:

- **Literature review**: This examines the current body of research on IoT security threats, including common attack vectors and case studies illustrating real-world breaches. The literature review also covers recent advancements and emerging trends in IoT security, identifying key gaps in the current solutions.
- **Methodology**: This includes description of data sources, threat assessment criteria, and the analytic framework used in assessing the security risks of IoT, including limitations of the chapter.
- **IoT security threats in smart environments**: This includes a detailed classification of security threats in IoT into device-specific threats, network-related threats, data privacy threats, and application-layer threats, with examples to illustrate each type.
- **IoT security vulnerability analysis**: This includes analysis of vulnerabilities for various IoT devices and environments, substantiated with case studies and statistical data.
- **Impact of security threats**: This chapter discusses the financial, privacy, and societal impacts of IoT security threats, underscoring the urgency of enhancing security measures.
- **Discussion and recommendations**: This includes best practices and future research directions to enhance security in IoT.
- **Conclusion**: This includes a summary of findings and a call for continued IoT security protocols and standards innovation.

This chapter tries to contribute to a safer, resilient IoT ecosystem by systematically identifying security threats and studying their implications. The findings and recommendations are helpful to researchers, developers, and policymakers to address the overwhelming challenges of securing the smart environment of IoT.

3.2 LITERATURE REVIEW

Integrating the Internet of Things technology into sectors like healthcare, industrial automation, smart homes, and city infrastructure has changed how people interact with technology and increased automation and efficiency. However, the widespread of IoT devices, which often have minimal security configuration, raises serious issues regarding security and privacy [6]. This literature review will provide an in-depth analysis of IoT security threats, examine the limitations of existing solutions, and explore emerging trends and potential future threats within smart environments.

3.2.1 Threat to IoT smart environment security today

IoT devices, therefore, differ from traditional computing systems in that they are often designed with minimal processing power, limited memory, and low-energy consumption requirements. These constraints often preclude complex security mechanisms, exposing them to cyberattacks that target their vulnerabilities [4]. Researchers have grouped IoT security threats into several primary categories.

3.2.1.1 Device-related threats

IoT devices generally come deployed with very little or outdated firmware, simple authentication mechanisms, and poor physical security [7]. One example is the Mirai Botnet attack that happened on a large scale, where weak passwords in IoT devices were used to compromise them and turn them into "zombies" that launched distributed denial-of-service attacks against critical internet infrastructure [8]. This attack sheds light on how IoT devices could be co-opted into botnets. Other device-related threats include the infection of malware, whereby malicious code is injected into the device's firmware, allowing attackers to gain control over device functions and spy on users. The enormity of this is amplified by the sheer number of IoT devices that rarely or poorly manage firmware updates since many IoT manufacturers prefer cost-effectiveness over security [9].

3.2.1.2 Network-related threats

IoT devices frequently communicate over wireless networks. The latter are generally not protected or are poorly safeguarded, thus exposing them to network-specific attacks. The common network-related threats include man-in-the-middle (MITM) attacks and packet sniffing [10]. In MITM attacks, an attacker will intercept communications between IoT devices and servers or other devices to gain unauthorized access to the transmitted data and potentially alter it [11]. The other standard attack technique involves packet sniffing, where network traffic is monitored to intercept sensitive information such as passwords, encryption keys, or even user data. IoT devices usually have weak encryption protocols due to the limitation in processing power, thus making it easy for an attacker to intercept or manipulate the transmitted data [12, 13].

3.2.1.3 Data privacy threats

IoT devices generally collect and process vast reams of personal data, such as health, location, and usage information [14]. This creates a massive threat to individual privacy. Unauthorized access may lead to a data breach, where sensitive information is exposed or stolen. Healthcare IoT devices such as wearable health monitors are particularly vulnerable since they accumulate sensitive data, and security measures to safeguard that information are usually feeble [15]. Insecure authentication mechanisms in these devices would thus allow attackers to access private data, compromising not only the user's privacy but also the confidentiality of the healthcare systems [16]. Insecure data storage and management practices are standard in IoT and tend to worsen data privacy threats. Data are often stored on devices or sent to cloud services using weak encryption or access control [15].

3.2.1.4 Application layer threats

IoT devices mainly connect with several third-party applications and extend the attack surface. Application-layer threats use software vulnerabilities to access APIs or outdated software components [17, 18]. Many vulnerabilities arise from unpatched software, weak encryption, and inadequate API security measures. Most IoT applications have no routine updates, so they become very vulnerable to attackers who exploit the common vulnerabilities to compromise users' data or even control the device in some cases. Besides, most applications are designed with minimum-security protocols for faster deployment and hence become easy targets for exploitation [17, 19].

3.2.2 Existing solutions and limitations

While meaningful research has been dedicated to developing security solutions for IoT, its unique characteristics make traditional security approaches less effective [20, 21]. Due to several obstacles, commonly recommended security measures, such as encryption, authentication, and firewall protection, are becoming increasingly difficult to enforce in IoT applications.

3.2.2.1 Encryption and authentication

Encrypting IoT data forms the basis for any form of cybersecurity, but in general, IoT devices do not possess enough computational power to execute traditional algorithms [22]. Lightweight encryption algorithms were proposed to replace the traditional ones because their processing requirements are lower; for example, Elliptic Curve Cryptography (ECC). However, lightweight encryption algorithms have been sparsely implemented despite having advantages due to high implementation costs and a lack of interoperability [23]. Authentication is another critical aspect of IoT security. Many IoT devices rely on either static passwords or default ones, which, in most cases, users do not change and, hence, are easily guessed or cracked [24]. The lack of multifactor authentication due to resource constraints further weakens IoT devices' defenses and exposes them to unauthorized access and brute-force attacks.

3.2.2.2 Network security solutions

Network segmentation and intrusion detection systems (IDS) can secure IoT devices at the network level. Network segmentation isolates IoT devices into contained networks that reduce the possibility of widespread compromise. However, maintaining network segmentation in IoT environments is challenging, as devices can often be mobile and dynamically connect to various networks. The other equally recommended measure is IDS [25]. However, it falls short in IoT due to the high false-positive rates coming from the dynamic nature of IoT traffic and unique communication protocols that differ from those of traditional computing environments. The current IDS tools need heavy tuning to adapt to the peculiarities of IoT data patterns, which are sometimes impossible because of resource limitations [25].

3.2.2.3 Software and firmware update

IoT devices require periodic updates since it is through updating the known vulnerabilities are addressed and the newest security patches are installed. However, most IoT devices do not support auto-update due to the characteristic resource-constrained or remote environment [26]. Without regular updates, these IoT devices will continue to be exposed to the known exploits that hackers are targeting. Moreover, the scale of IoT deployment makes firmware updates even more difficult, as manually updating each device in a vast network of diverse devices is impractical. In addition, most manufacturers stop maintaining their IoT devices in due course, resulting in an ever-widening security gap as the devices age [26].

3.2.3 Emerging trends and threats

The advancement of the Internet of Things (IoT) technology continues introducing new security challenges, especially as IoT devices integrate with other cutting-edge technologies such as artificial intelligence, edge computing, and blockchain. Each of these technologies brings both security enhancements and new potential attack vectors [27].

3.2.3.1 AI-driven attacks and defenses

The capabilities of artificial intelligence power the security of IoT devices, mainly automating the detection of threats and anomalies and the mechanisms for responding to them. AI-driven defenses are now executing machine-learning algorithms that can analyze network traffic, identify patterns indicative of attacks, and respond autonomously [28]. However, attackers can also weaponize AI by developing adaptive and sophisticated malware that evades detection. For example, attackers can leverage AI to analyze and predict IoT device behaviors, enabling malware to change tactics dynamically to evade detection [29]. This creates an ongoing "arms race" where both attackers and defenders try to gain the upper hand using AI.

3.2.3.2 Edge computing vulnerabilities

In edge computing, data is processed nearer to its source—IoT devices—rather than sent to centralized cloud servers. Although edge computing lowers latency and reduces bandwidth, it creates new vulnerabilities. Although numerously deployed, edge nodes—devices or servers placed near data sources—are much less secure than traditional data centers and become exposed to different threats, such as data manipulation, unauthorized access, and distributed denial-of-service attacks. Because edge nodes do not have strong security features of centralized architectures, attackers can use these nodes to penetrate IoT networks and manipulate data or disturb IoT functions [29].

3.2.3.3 Blockchain-based security solutions

Blockchain technology, a decentralized and tamper-proof ledger, has been explored to a certain extent in providing IoT security solutions [30]. It provides the fundamental framework for interactions between devices, authentication, and data integrity by storing the records in distributed ledger technology, which is immutable and verifiable. However, high computational requirements and power consumption raise significant challenges for implementing blockchain on resource-constrained IoT devices. Moreover, the scalability of blockchain itself is a considerable concern, as storage and processing may demand more than what is usually provided by IoT infrastructure in large-scale deployments [31].

3.2.4 Summary of literature findings

It has been well noted in the literature that, although considerable strides have been made to understand and address IoT security threats, challenges persist due to the resource constraints of IoT devices and the unique characteristics of IoT environments [32]. Traditional security measures, such as encryption and firewalls, are usually not enough for IoT, while the emerging solutions in blockchain and AI are still in the experimental stage, with various limitations to be addressed before being adopted on a large scale [33]. Further, new technologies like AI and edge computing integration with IoT will make the security landscape continue to evolve—both in finding new capabilities and new risks.

The review underlines the need for a multilayered, multifaceted IoT security approach on device-specific, network-level, and data privacy controls. The following sections of the chapter will explore these findings in depth, giving an in-depth look into IoT security threats and their impacts and presenting good practices to enhance the security of IoT applications in different smart environments.

3.3 METHODOLOGY

The main objective of this research is to identify, categorize, and analyze the security threats in IoT smart environments and assess effective measures to counter them. This section describes the methodology adopted for data collection, its analysis, and the design of threat mitigation models tailored for IoT environments. A mixed-method approach included quantitative analysis of security incidents and qualitative evaluations of threat types, their impact on system functionality, and potential solutions.

3.3.1 Research design

This research adopts a descriptive and analytical design to assess IoT security threats comprehensively. The methodology of the chapter includes the following steps:

1. **Literature review and framework development**: A wide-range literature review was performed to identify common IoT threats and evaluate existing solutions.
2. **Data collection**: Data on recent security incidents and vulnerabilities in IoT systems was gathered from threat databases, industry reports, and case studies.
3. **Threat categorization**: Threats were grouped into different types (device-level, network-level, data privacy, and application-layer threats).
4. **Analysis and modeling**: Statistical analysis helped define the prevalence and impact of each type of threat, while qualitative analysis put each threat's root cause into context.
5. **Evaluation of alternative solutions**: Security measures are evaluated for effectiveness and shortcomings.

These stages are visualized in Table 3.1.

3.3.2 Data collection

Data collection was tailored to gather quantitative and qualitative information from various sources, ensuring a comprehensive dataset for analyzing

Table 3.1 Research methodology stages in IoT security threat analysis

Stage	*Description*
Literature Review	Comprehensive review of IoT security threats and existing solutions
Data Collection	Gathering data from IoT threat databases, case studies, and industry reports
Threat Categorization	Grouping threats by type to provide a structured understanding of IoT vulnerabilities
Analysis and Modeling	Conducting statistical and qualitative analyses of threat frequency and impact
Evaluation of Solutions	Assessing the effectiveness of existing security measures

Table 3.2 Data sources and purposes for IoT threat analysis

Source	*Data type*	*Purpose*
Threat Intelligence Databases	Quantitative (vulnerability data)	Identifying common vulnerabilities in IoT devices
Industry Reports and Case Studies	Qualitative	Insight into the impact and consequences of IoT security breaches
Survey of IoT Practitioners	Qualitative	Expert perspectives on current and emerging threats
Device and Network Logs	Quantitative and Qualitative	Traffic and threat pattern analysis

IoT security threats. Table 3.2 shows data sources and purposes for IoT threat analysis. The data sources included:

1. **Threat intelligence databases**: The National Vulnerability Database (NVD) and Common Vulnerabilities and Exposures (CVE), among the many available platforms, provided real-world examples of vulnerabilities in IoT devices.
2. **Industry reports and case studies**: Reports from cybersecurity firms and case studies of IoT security breaches offered valuable insights into threat patterns and the consequences of inadequate security measures.
3. **IoT security practitioner survey**: An IoT security practitioner survey was conducted to garner an expert view on current and future threats and the effectiveness of currently existing security solutions.
4. **Device and network logs**: Anonymized logs provided by IoT networks under investigation were analyzed to study traffic patterns, intrusion attempts, and vulnerabilities.

3.3.3 Data analysis techniques

Data was analyzed using a mixed-method approach incorporating quantitative statistical techniques and qualitative thematic analysis.

3.3.4 Statistical analysis

Quantitative data analysis involved calculating the frequency of each type of IoT threat using descriptive statistics and trend analysis. This helped us understand specific threats' prevalence and growth trends, such as device-specific attacks or network vulnerabilities.

- Frequency distribution: The researcher formed a frequency distribution of threat occurrences that presented the most current type of threats in recent years.
- Trend analysis: Trend analysis was used to view temporal changes in the prevalence of threats.

Figure 3.1 represents the distribution of IoT security threats over recent years, highlighting the types with increased or decreased frequency.

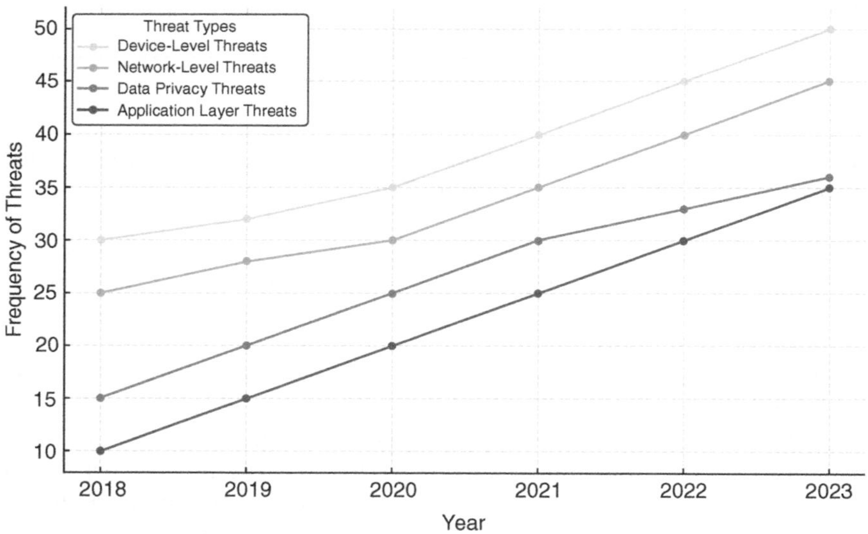

Figure 3.1 Frequency of IoT threat types (2018–2023).

Table 3.3 IoT threat categorization model

Category	*Characteristics*	*Example threats*
Device-Level Threats	Exploit weaknesses in device firmware/hardware	Firmware manipulation, malware
Network-Level Threats	Exploit network communication vulnerabilities	Man-in-the-middle, packet sniffing
Data Privacy Threats	Unauthorized access to sensitive data	Data breaches, unauthorized access
Application Layer Threats	Exploit software/API vulnerabilities	API abuse, code injection

3.3.4.1 Qualitative analysis

A thematic analysis evaluated qualitative data, including case studies and survey responses. Key themes were extracted regarding the impact of each threat, the methods used to exploit IoT vulnerabilities and the perceived effectiveness of current security measures.

3.3.4.2 Threat categorization model

A threat categorization model was created based on data analysis, grouping threats by characteristics such as attack vectors, impact levels, and potential mitigation strategies. Each category was defined and analyzed with representative examples, as shown in Table 3.3.

3.3.5 Development of IoT security threat models

From the data analyzed, threat models were designed to express how threats propagate in IoT environments. The models included possible attack vectors, potential impacts, and common security difficulties in preventing such attacks.

1. **Threat propagation model**: This model illustrates how an attack on an IoT device spreads throughout a network. For instance, malware infection on a single device can compromise network integrity, allowing further attacks on additional devices.
2. **Impact assessment model**: The impact assessment model was designed to classify threats according to their level of impact, ranging from minor disruptions in service to severe data loss or breaches in system control. This helped further prioritize threats by their potential harm to the IoT environment.
3. **Threat response model**: This model outlines the necessary steps to mitigate each category of threat, suggesting specific security protocols, software updates, or network configurations to strengthen IoT security.

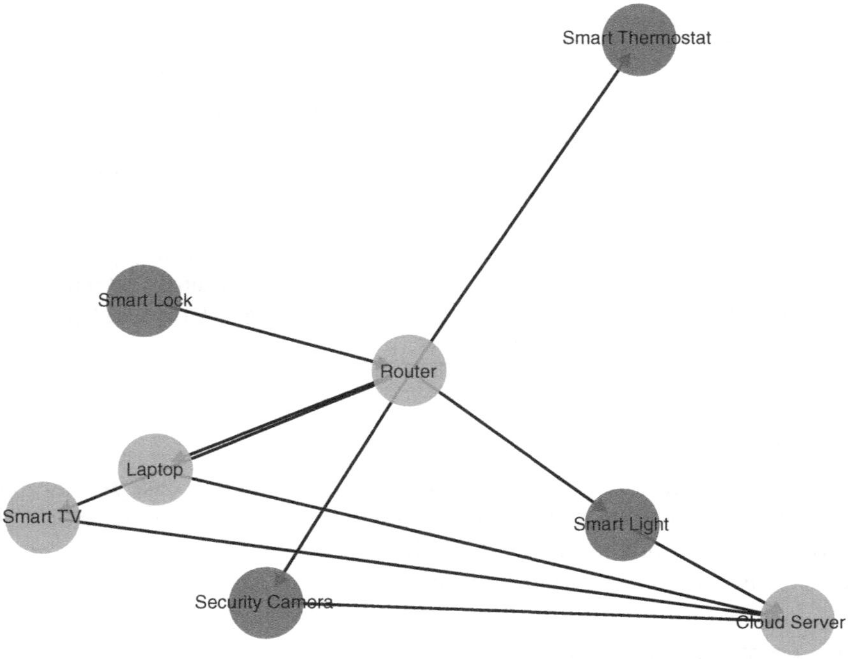

Figure 3.2 Threat Propagation Pathways in IoT Networks.

Figure 3.2 illustrates typical pathways for malware propagation across IoT devices, highlighting weak points in network security.

3.3.6 Review of the current security controls

Common IoT security solutions were evaluated alongside threat categorization and analysis using expert surveys and literature data. The evaluation considers these security solutions' effectiveness, scalability, and suitability for resource-constrained IoT environments. Table 3.4 provides an evaluation of IoT security measures.

1. **Mechanisms of encryption**: Lightweight encryption mechanisms, such as ECC, were analyzed regarding their feasibility and effectiveness within IoT environments.
2. **Authentication techniques**: The effectiveness of multifactor authentication techniques in reducing unauthorized access to IoT devices was analyzed.
3. **Intrusion detection systems**: The feasibility of IDS in IoT was investigated by examining the tradeoff between detection accuracy and false-positive rates in a dynamic IoT network.

Table 3.4 Evaluation of IoT Security Measures

Security measure	*Effectiveness*	*Limitations*
Lightweight Encryption	High for data protection	Limited by device processing capabilities
Multifactor Authentication	High for access control	Implementation challenges in low-power devices
Intrusion Detection System	Moderate for network security	A high false-positive rate requires customization

3.3.7 Limitations of the methodology

Though the methodology provides a structured approach to identifying and analyzing threats in IoT security, it has some limitations.

1. **Data access constraints**: The access to proprietary data from some IoT systems and networks may be limited, limiting the threat analysis.
2. **The dynamic nature of IoT technology**: The rapidly changing landscape of IoT may bring new threats that are not included in the present analysis, and hence, the findings may lose their relevance over time.
3. **Generalizability of findings**: The findings in one IoT environment, such as healthcare or industrial, may not directly apply to other sectors because the analysis needs further customization for industry-specific threats.

The above methodology supplies a structured approach toward understanding and addressing IoT security threats in bright environments. This research utilizes data collection, threat categorization, model development, and evaluation to present a holistic view of the IoT threat landscape. Findings from this methodology inform the following sections, which detail best practices for security, potential solutions, and future research directions in IoT security.

3.4 TYPES OF SECURITY THREATS IN IoT SMART ENVIRONMENTS

IoT smart environments comprise interconnected devices, networks, and applications that facilitate communication, automation, and data exchange across various applications. However, these environments are highly susceptible to multiple security threats due to their networked nature and varying degrees of built-in security. The following sections identify and describe the main security threats affecting IoT smart environments: device-level, network-level, data privacy, and application-layer threats.

3.4.1 Device-level threats

Device-level threats target the hardware and firmware of IoT devices. Since many IoT devices have limited processing power and storage capacity, they often lack sophisticated security features, making them vulnerable to malware, firmware manipulation, and physical tampering. Table 3.5 provides types of device-level threats and their impact on IoT environments.

- **Malware and ransomware**: Malware increasingly targets IoT devices, which can infect and spread across networks. Malware compromises device functionality and can lead to data loss or malfunction. Ransomware, a specific type of malware, locks users out of devices or encrypts data, requiring a ransom to restore access.
- **Firmware manipulation**: Attackers can exploit vulnerabilities in device firmware, manipulating it to gain unauthorized control or to introduce backdoors for ongoing access. Firmware attacks can be challenging to detect and can persist even after software updates.
- **Physical tampering**: IoT devices, especially those deployed in unsecured locations, are susceptible to physical attacks. Attackers can directly access device components and tamper with the hardware to alter functionality or extract sensitive information.

Figure 3.3 shows the relative frequency of each device-level threat in each IoT environment, illustrating which threats are more prevalent based on real-world data from threat intelligence databases.

3.4.2 Network-level threats

Network-level threats target communication channels between IoT devices. They exploit protocols, encryption, or configuration vulnerabilities. Network attacks can compromise the integrity, availability, and confidentiality of data transmitted across IoT networks. Table 3.6 shows network-level threats in IoT smart environments.

Table 3.5 Types of device-level threats and their impact on IoT environments

Threat	*Description*	*Impact*
Malware and Ransomware	Infects devices, disrupt functionality, demands a ransom	Data loss, system downtime, financial loss
Firmware Manipulation	Alters firmware to gain control or create backdoors	Persistent unauthorized access
Physical Tampering	Direct access to device components for manipulation	Device malfunction, data extraction

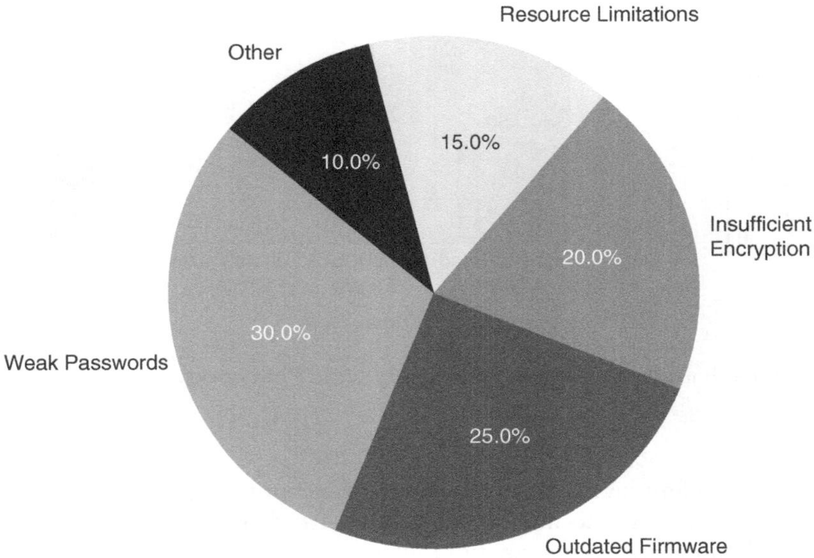

Figure 3.3 Distribution of device-level threats in IoT systems.

Table 3.6 Network-level threats in IoT smart environments

Threat	*Description*	*Impact*
Man-in-the-Middle (MITM)	Intercepts device communication for data capture or injection	Data compromise, unauthorized control
Denial of Service (DoS/DDoS)	Overloads network with requests, causing downtime	Service disruption, potential system crashes
Packet Sniffing	Captures data packets to extract sensitive information	Exposure of credentials, data leakage

1. **Man-in-the-Middle (MITM) Attacks**: In an MITM attack, an attacker will intercept communication between devices to capture sensitive information or inject malicious data. Without strong encryption, MITM attacks can easily compromise data integrity.
2. **Denial of Service (DoS) and Distributed Denial of Service (DDoS)**: These attacks overload IoT networks or specific devices with excessive requests, causing service disruption or shutdown. The sheer volume of IoT devices in networks can amplify the effect of DDoS attacks.

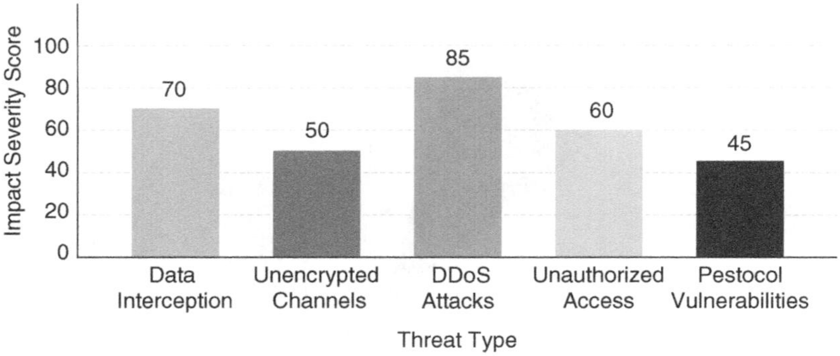

Figure 3.4 Impact of network-level threats on IoT networks.

3. **Packet Sniffing**: Packet sniffing involves capturing packets transmitted over networks to monitor data flow. Attackers use packet sniffing to obtain sensitive information, including authentication credentials or unencrypted data.

Figure 3.4 illustrates the impact of each network-level threat by quantifying potential downtime, data loss, or service interruptions and comparing each threat's severity of effects on IoT operations.

3.4.3 Data privacy threats

Data privacy threats in IoT environments mean unauthorized access to sensitive information and impact personal and critical operational data. Considering the volume of data generated by IoT devices, the privacy risks are drastically high when such devices collect and transmit sensitive information over networks. Table 3.7 provides types of data privacy threats in IoT environments.

1. **Unauthorized data access**: Incompetent access control may allow an attacker to access sensitive data on or sent via the IoT device. This may be used for identity theft, breaching regulations, and losing consumer confidence.
2. **Data breaches**: IoT environment breaches may expose enormous quantities of personal and operational data as attackers target central data repositories, cloud storage, or on-device storage.
3. **Insufficient data encryption**: Many IoT devices lack adequate encryption protocols for data in transit or at rest, making data easily accessible to attackers who intercept or physically access devices.

Table 3.7 Types of data privacy threats in IoT environments

Threat	*Description*	*Impact*
Unauthorized Data Access	Weak access control allows unauthorized data access	Identity theft, regulatory non-compliance
Data Breaches	Attacks targeting data storage to extract large datasets	Loss of sensitive information, reputational damage
Insufficient Data Encryption	Lack of strong encryption exposes data during transmission	Data interception, unauthorized data access

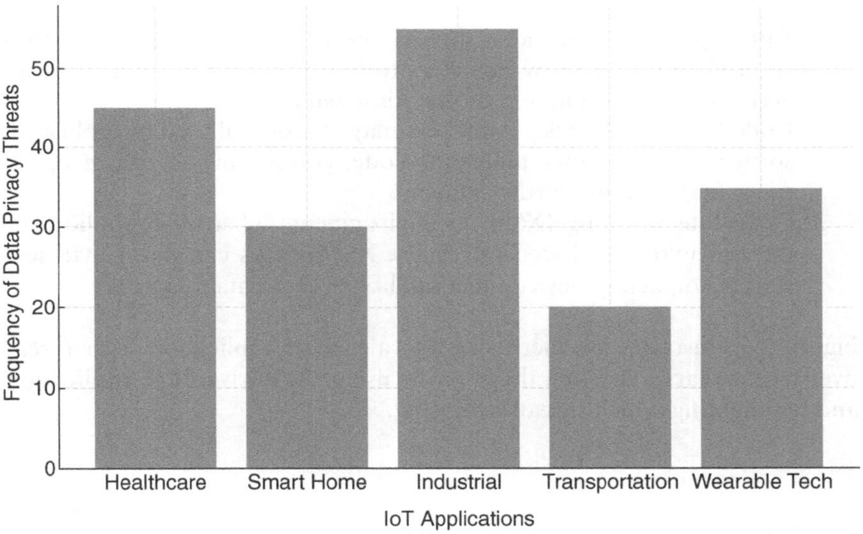

Figure 3.5 Frequency of data privacy threats in IoT applications.

Figure 3.5 displays the occurrence of data privacy threats across different IoT applications, highlighting which applications (e.g., healthcare, smart home, industrial) are more prone to specific data privacy threats.

3.4.4 Application layer threats

Application-layer threats exploit vulnerabilities in the software, APIs, or applications communicating with IoT devices. To make matters worse, many IoT devices depend upon applications for their remote management and data processing; hence, applications become the primary targets of attackers. Table 3.8 shows the application-layer threats affecting IoT environments.

Table 3.8 Application layer threats affecting IoT environments

Threat	*Description*	*Impact*
API Exploits	Vulnerabilities in APIs allow unauthorized access	Data manipulation, unauthorized device control
Code Injection Attacks	Malicious code injected to exploit application weaknesses	Operational disruption, potential data compromise
Cross-Site Scripting (XSS)	Injects scripts into applications that users interact with	Data corruption, user data exposure

1. **API Exploits**: IoT devices often connect to apps using APIs. Poorly secured APIs can allow attackers to bypass authentication, allowing access to or manipulation of unauthorized data.
2. **Code Injection Attacks**: Attackers may exploit vulnerable application software by injecting malicious code, compromising device operations, and enabling further attacks.
3. **Cross-Site Scripting (XSS)**: In environments where IoT applications interact with web-based interfaces, XSS attacks can inject malicious scripts, impacting device functionality or user data integrity.

Figure 3.6 illustrates the increasing prevalence of application layer threats over recent years, reflecting the growing use of API-driven IoT applications and highlighting which threats are rising.

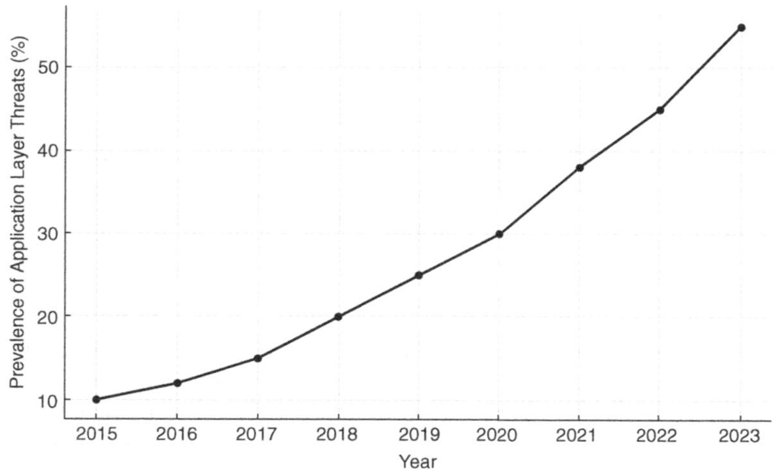

Figure 3.6 Prevalence of application layer threats in IoT smart environments.

3.4.5 Summary of threat impact on IoT environments

Each threat type impacts the core information security principles in IoT environments: confidentiality, integrity, and availability (CIA). Device-level threats often hit availability, network-level threats compromise data integrity, data privacy threats undermine confidentiality, and application-layer threats pose risks across all three pillars of the CIA triad. Table 3.9 shows a summary of threat types and their security impacts on IoT.

Figure 3.7 illustrates each threat type's impact on confidentiality, integrity, and availability, highlighting which threats are critical in specific IoT

Table 3.9 Summary of threat types and their security impacts on IoT

Threat type	*Primary security impact*	*Affected CIA principle*
Device-Level Threats	Device functionality, system uptime	Availability
Network-Level Threats	Data transmission integrity	Integrity
Data Privacy Threats	Exposure to sensitive information	Confidentiality
Application Layer Threats	Software integrity, user data safety	Confidentiality, Integrity, Availability

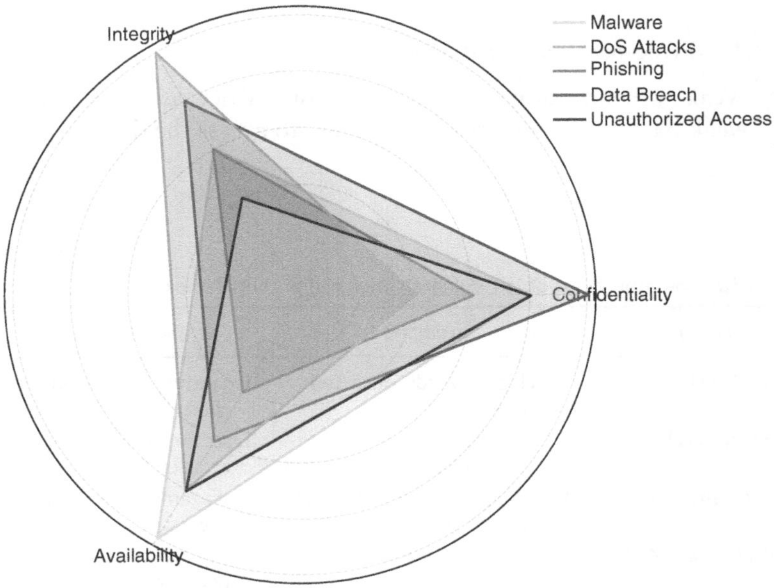

Figure 3.7 CIA impact of different threat types in IoT smart environments.

settings. Knowing the kinds of security threats in IoT smart environments will help to narrow down and enhance the effectiveness of mitigation strategies. From device-level vulnerabilities to application-layer exploits, each type of threat presents unique challenges in maintaining IoT's secure operations. By identifying those threats and their impacts on CIA principles, IoT system designers and security experts could better protect against possible risks and build a resilient and trustworthy IoT ecosystem.

3.5 IoT SECURITY VULNERABILITIES ANALYSIS

The different design, deployment, and management weaknesses of devices and networks in IoT environments lead to IoT security vulnerabilities. One of the significant causes of unauthorized access, data breaches, and system disruptions in the IoT ecosystem is vulnerabilities. In this section, we present the categorization of common IoT security vulnerabilities, analyze them, discuss their root causes, and present the level of their impact on security.

3.5.1 Common IoT vulnerability types

This report will focus on the four most common types of IoT device vulnerabilities: device-level, network-level, data-level, and application and API. Table 3.10 shows common types of IoT vulnerabilities and potential impacts.

3.5.1.1 Device-level vulnerabilities

- Weak authentication mechanisms: Most IoT devices rely on default or weak passwords that attackers can use to access the device.

Table 3.10 Common types of IoT vulnerabilities and potential impacts

Vulnerability type	*Examples*	*Potential impacts*
Device-Level	Weak passwords, outdated firmware	Unauthorized access, data theft
Network-Level	Unencrypted channels, vulnerable protocols	Data interception, MITM attacks
Data Transmission	Insufficient integrity checks, insecure storage	Data tampering, unauthorized access
Application and API	Insecure APIs, weak access control	Data manipulation, unauthorized control

- Insecure Firmware: IoT devices usually run outdated firmware with known vulnerabilities. Infrequent firmware updating allows several devices to be exposed to known exploits.
- Resource Constraints: Due to limited computational resources, many IoT devices cannot support complex security protocols, which increases their susceptibility to attacks.

3.5.1.2 *Network-level vulnerabilities*

- Unencrypted Communication Channels: Many IoT devices communicate over unencrypted channels, exposing sensitive information to interception.
- Insecure Communication Protocols: The standard protocols in IoT, when used in an insecure way, are vulnerable to attacks so that unauthorized access might be given to any communication sent by the device.

3.5.1.3 *Data transmission vulnerabilities*

- Data Integrity Checks: Some IoT devices are vulnerable to data corruption and injection attacks due to a lack of proper data integrity checks.
- Insecure Data Storage: IoT devices may store data locally without encryption or access controls, exposing sensitive data to unauthorized access.

3.5.2 Application and API vulnerabilities

- Insecure APIs: APIs are the standard interface for communication between applications and IoT devices. Unsecured APIs can offer an entry to an attacker, who can now manipulate data or control devices by leveraging the API.
- Lack of Access Control Measures: Weak application or API access controls allow unauthorized users to exploit devices or data.

Figure 3.8 illustrates the distribution of these vulnerability types across IoT environments, showing which types are most prevalent.

3.6 IoT SECURITY VULNERABILITIES: ROOT CAUSES

The root causes of vulnerabilities in IoT are usually vast and can be attributed to factors ranging from device limitations to inadequate security practices in development and deployment. Understanding these causes provides

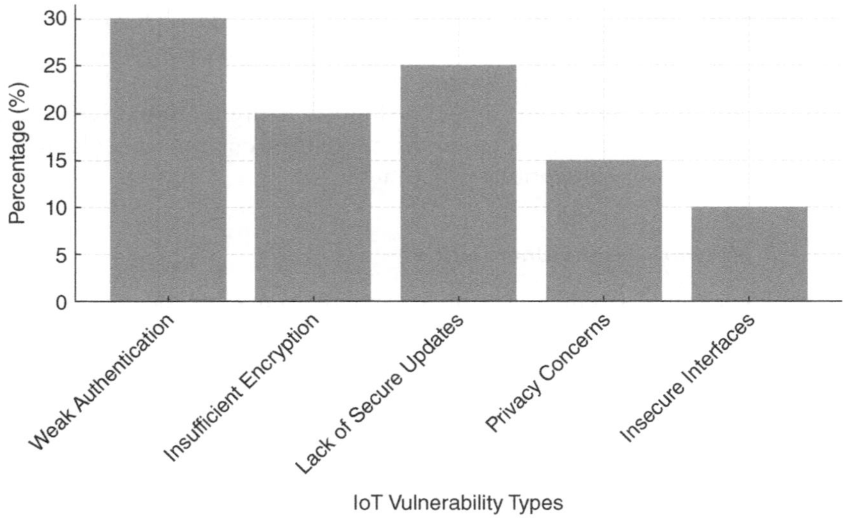

Figure 3.8 Distribution of IoT vulnerability types.

Table 3.11 Root causes of IoT security vulnerabilities and their impacts

Root cause	*Description*	*Impact on vulnerabilities*
Limited Computational Resources	Constrained storage and processing power limit security features	Weak encryption, inadequate authentication
Short Product Development Cycles	Rapid market demands limit security testing	Unpatched vulnerabilities, insufficient testing
Inadequate Security Standards	The lack of universal standards allows manufacturers to deprioritize security.	Weak industry-wide security practices
Complexity of IoT Networks	Varying device protocols and compatibility issues	Interoperability issues, unaddressed gaps

insight into how vulnerabilities arise and persist. Table 3.11 shows root causes of IoT security vulnerabilities and their impacts.

1. **Limited computational resources**: IoT devices are designed with constrained efficiency in mind, which usually translates to smaller storage, processing power, and energy resources. The latter may not be strong enough to support the more resource-intensive security measures, such as sophisticated encryption and repeated checks for authentication.

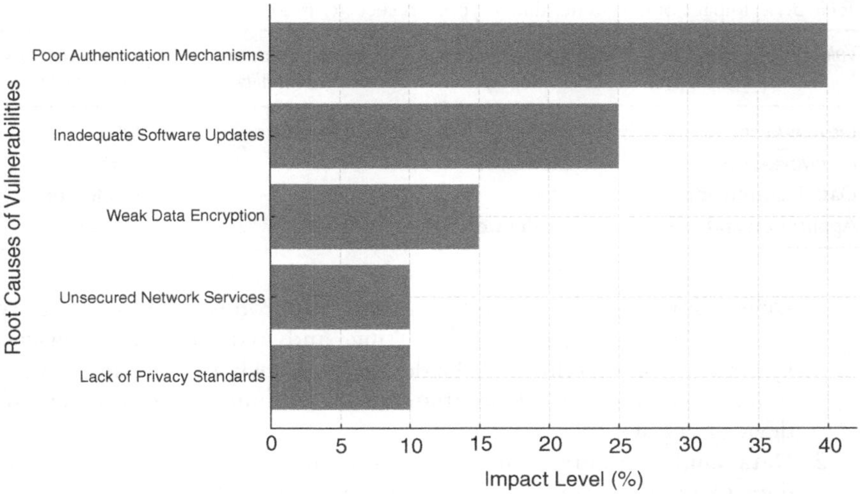

Figure 3.9 Root causes of vulnerabilities in IoT devices.

2. **Short product development cycles**: Many IoT products are rushed to market to address demand. This can lead to a lack of thorough security testing, meaning devices are sent out with exploitable vulnerabilities. Security is often considered an add-on feature, not a core requirement, predisposing a device to vulnerabilities.
3. **Inadequate security standards and regulations**: There is a lack of universally enforced security standards for IoT devices. Without stringent regulations, manufacturers may not be motivated to prioritize security, creating devices that are easily compromised.
4. **Complexity of IoT networks**: The IoT environment usually contains many devices from various manufacturers, and the security features of these devices are diverse. This diversity creates compatibility issues and could result in security holes when different devices communicate on the same network.

Figure 3.9 represents the impact level of each root cause on IoT vulnerabilities, showing which root causes are most critical for industry consideration.

3.6.1 Impact assessment of IoT vulnerabilities

The vulnerability of IoT has three significant metrics to assess the impact: Device Integrity, Data Confidentiality, and Network Availability. Table 3.12 shows the impact of IoT vulnerability types on security metrics. Every vulnerability type affects these metrics differently, as shown in Table 3.12.

Table 3.12 Impact of IoT vulnerability types on security metrics

Vulnerability type	*Impact on device integrity*	*Impact on data confidentiality*	*Impact on network availability*
Device-Level	High	Moderate	Low
Network-Level	Moderate	High	High
Data Transmission	Low	High	Moderate
Application and API	Moderate	High	Moderate

1. **Device integrity**: Vulnerabilities that compromise device integrity can lead to access, control hijacking, and malfunctioning devices. Generally, vulnerabilities at the device level, such as weak authentication and firmware update insatiateness, contribute the most directly to device integrity.
2. **Data confidentiality**: Vulnerabilities in data confidentiality expose data to unauthorized access and leakage. Data transmission vulnerabilities expose sensitive information, such as unencrypted channels and insufficient data integrity checks.
3. **Availability**: Some vulnerabilities impact network availability and cause service disruptions and denial-of-service conditions. The usual cause of network unavailability is network-level vulnerabilities that involve insecure protocols and unencrypted communication.

Figure 3.10 shows how each type of vulnerability impacts device integrity, data confidentiality, and network availability, highlighting areas requiring focused security measures.

3.6.2 Analysis of vulnerability patterns

Analyzing the patterns in IoT vulnerabilities reveals trends that can help predict future threats and guide the development of preventive measures. Table 3.13 shows IoT vulnerability patterns and implications.

1. **Increase in network-level vulnerabilities**: Network-level vulnerabilities are rising, with IoT ecosystems becoming more interconnected. This trend will continue as more devices connect via networks and share data without strong encryption.
2. **Emerging threats in API security**: As IoT devices increasingly rely on API security, it has become a significant concern for PIs for functionality and integration with other systems. Insecure APIs introduce risks for unauthorized data access and control over devices.
3. **Vulnerability persistence in firmware**: Firmware-related vulnerabilities persist due to the lack of regular updates and patches. Many IoT devices remain vulnerable to known exploits simply because their firmware is outdated or unsupported.

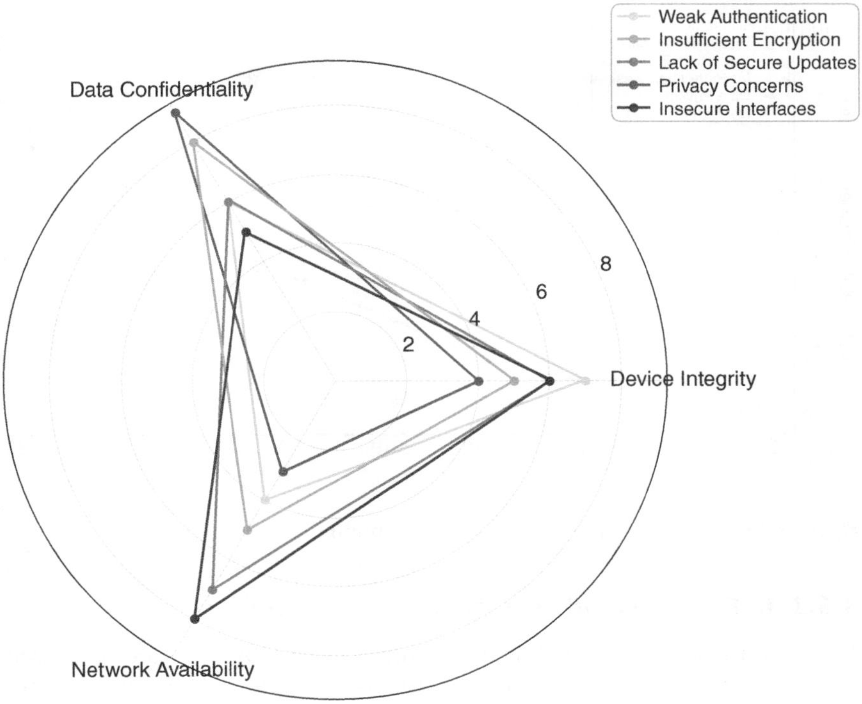

Figure 3.10 Vulnerability impact across security metrics.

Table 3.13 IoT vulnerability patterns and implications

Pattern	*Description*	*Implications for future security*
Increase in Network Vulnerabilities	Growing interconnectivity raises risks of interception and MITM attacks	Emphasis on network security and encryption
API Security Threats	Increasing API reliance introduces new avenues for attacks	Necessity for secure API development practices
Firmware Vulnerability Persistence	Lack of updates leaves devices susceptible to known exploits	Regular firmware updates and patching are essential

Figure 3.11 shows the growth of each vulnerability type over time, allowing readers to observe which vulnerabilities are becoming more prominent in IoT environments.

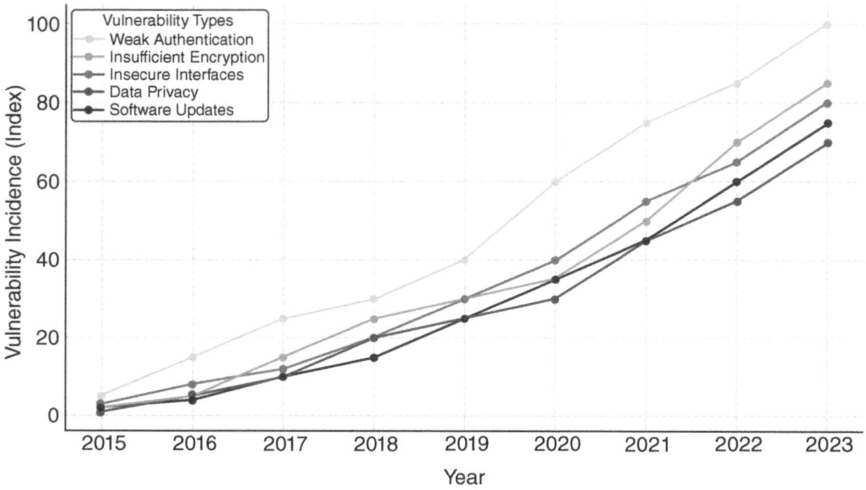

Figure 3.11 Vulnerability trends in IoT environments.

3.6.3 IoT vulnerabilities: Mitigation strategies

Based on the analysis of IoT vulnerabilities, mitigation strategies are essential to deal with security weaknesses effectively. Table 3.14 provides recommended mitigation strategies for IoT vulnerabilities.

1. **Strong authentication and access control**: To curtail unauthorized access to IoT devices, multifactor authentication, secure password practices, and role-based access controls are recommended.
2. **Regular firmware and software updates**: Like all other IT systems, IoT devices need to be updated regularly to patch vulnerabilities and address any known exploited conditions. Mechanisms for automated updates can be implemented.

Table 3.14 Recommended mitigation strategies for IoT vulnerabilities

Mitigation strategy	*Description*	*Targeted vulnerabilities*
Strong Authentication and Access Control	Implement multifactor authentication and access controls	Device-Level, Application and API
Regular Firmware and Software Updates	Apply automated updates to fix known vulnerabilities	Device-Level, Network-Level
Data Encryption and Integrity Verification	Encrypt data at rest and in transit; integrity checks	Data Transmission, Network-Level
Secure API Development	Use secure APIs with firm access control	Application and API

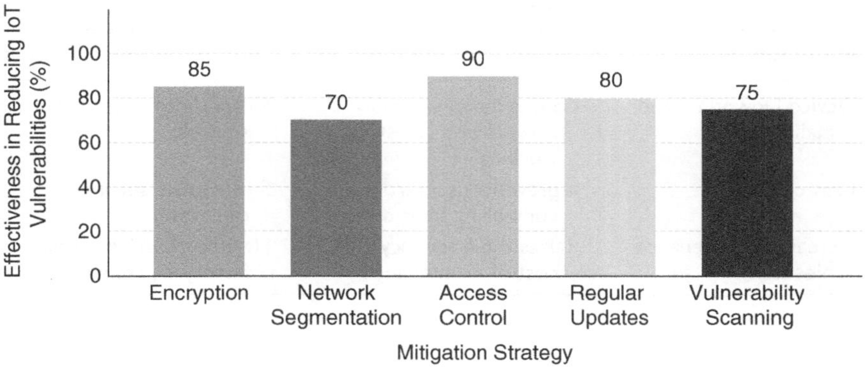

Figure 3.12 Effectiveness of mitigation strategies.

3. **Data can be encrypted for confidentiality**: Integrity checks shall also be carried out to ensure no unauthorized access or tampering with the data occurs.
4. **Secure API development**: API security can be enforced by requiring authentication and authorization, coupled with regular security assessments, to prevent unauthorized access to data and control of IoT devices.

Figure 3.12 compares the effectiveness of each mitigation strategy in reducing various types of IoT vulnerabilities, showing which strategies offer the most comprehensive protection.

Analyzing IoT security vulnerabilities brings significant gaps in device-level, network-level, data transmission, and application security. Understanding these vulnerabilities and their root causes and trends is instrumental in developing focused mitigation strategies for better protection in IoT ecosystems. As IoT networks continue to grow, strong security measures must be implemented to prevent future security incidents that could permanently dent the safety and integrity of IoT ecosystems.

3.7 IMPACT OF SECURITY THREATS

In smart IoT environments, security threats affect individual devices' functionality and the excellent network of connected systems. These could be minor inconveniences or massive security breaches, with one of the possible consequences being unauthorized access to data or substantial financial losses. Some of the domains through which the influence of these security threats is shown include device functionality, network performance, data

Table 3.15 Impact on device functionality due to IoT security threats

Impact type	*Description*	*Examples of affected devices*
Device Downtime and Malfunction	Causes devices to slow down, freeze, or stop working	Smart thermostats, industrial sensors
Loss of Control	It prevents users from controlling their devices	Smart locks, security cameras
Increased Maintenance Needs	Raises the frequency and cost of maintenance	Healthcare IoT devices, factory equipment

privacy, financial implication, and user trust. Table 3.15 provides impact on device functionality due to IoT security threats.

3.7.1 Impact on device functionality

Such security threats, especially those against the devices themselves, can cause severe degradation in device functionality, malfunction, or loss of control over the device.

- **Device downtime and malfunction**: Malware infection and denial-of-service (DoS) attacks often cause devices to slow down, freeze, or become unresponsive. In an interrelated IoT network, the failure of one device could cascade, impacting other devices that depend on it.
- **Loss of control**: Threats such as ransomware and firmware manipulation allow attackers to control devices, potentially locking legitimate users or causing devices to perform unintended actions.
- **Increased maintenance**: A compromised device usually requires immediate maintenance or replacement, disrupting IoT operations and increasing maintenance costs, especially in critical applications like industrial IoT.

Figure 3.13 illustrates the percentage of devices experiencing downtime, loss of control, and increased maintenance due to security threats, highlighting the most vulnerable IoT devices.

3.7.2 Impact on network performance

Network-level threats, such as DoS attacks and packet sniffing, directly impact the performance of IoT networks by breaching data flow, bandwidth, and network availability.

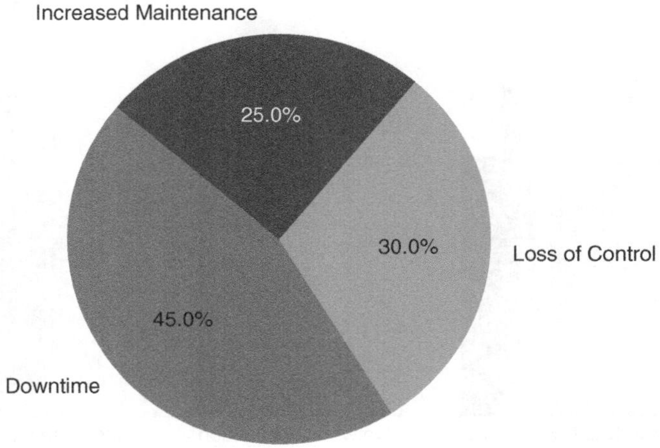

Figure 3.13 Device functionality impact from security threats.

- **Bandwidth exhaustion**: In DDoS attacks, an influx of malicious traffic can overwhelm the IoT network, consuming all available bandwidth and leaving devices unable to communicate effectively. This bandwidth exhaustion severely limits system functionality.
- **Network congestion**: Network congestion due to packet sniffing or unauthorized traffic slows communication among IoT devices and, in turn, affects time-critical applications such as autonomous vehicles and healthcare monitoring systems.
- **Latency**: High latency introduced by network attacks diminishes response times, which can be critical in applications such as smart grids or emergency response systems, where delays can result in a failed service. Table 3.16 provides data on network performance impact due to IoT security threats

Table 3.16 Network performance impact due to IoT security threats

Impact type	*Description*	*Examples of affected systems*
Bandwidth Exhaustion	Limits device communication and system functionality	Smart homes, industrial IoT
Network Congestion	Slows down data flow, affecting time-sensitive apps	Healthcare monitoring, autonomous vehicles
Increased Latency	Causes delays, critical in rapid-response settings	Emergency response systems, smart grids

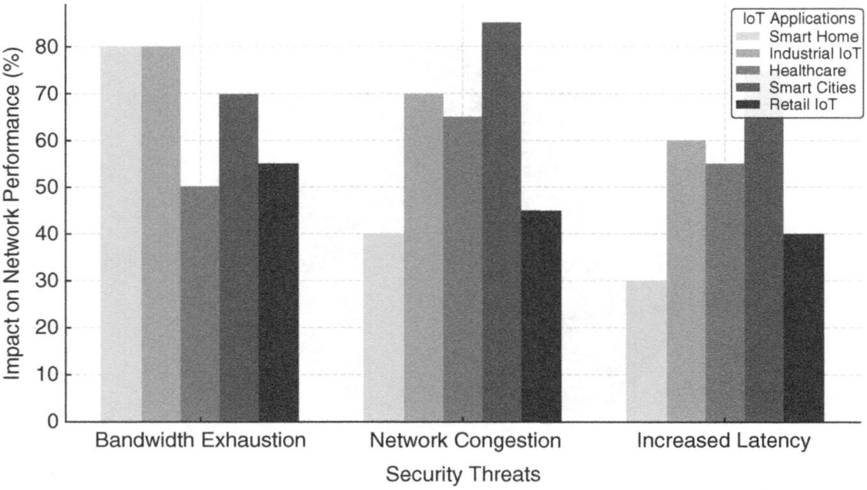

Figure 3.14 Network performance impact from security threats.

Figure 3.14 displays the impact of bandwidth exhaustion, network congestion, and increased latency across different IoT applications, emphasizing where network performance is most affected.

3.7.3 Impact on data privacy

IoT systems often handle large volumes of sensitive data, making them attractive targets for data privacy threats. The consequences of compromised data privacy may be dire for individuals and organizations. Table 3.17 shows the data privacy impact due to IoT security threats.

- **Exposure of sensitive information**: Such threats as unauthorized access and data breaches may expose personal and operational data, resulting in identity theft, regulatory penalties, and reputational damage.

Table 3.17 Data privacy impact due to IoT security threats

Impact type	*Description*	*Examples of sensitive data*
Exposure to Sensitive Info	Unauthorized access to personal or business data	Personal health data, business operational data
Loss of Data Confidentiality	Data is intercepted and accessed by unauthorized users	Financial information, control system data
Data Integrity Compromise	Data is altered or corrupted, leading to unreliable results	Sensor readings, patient health records

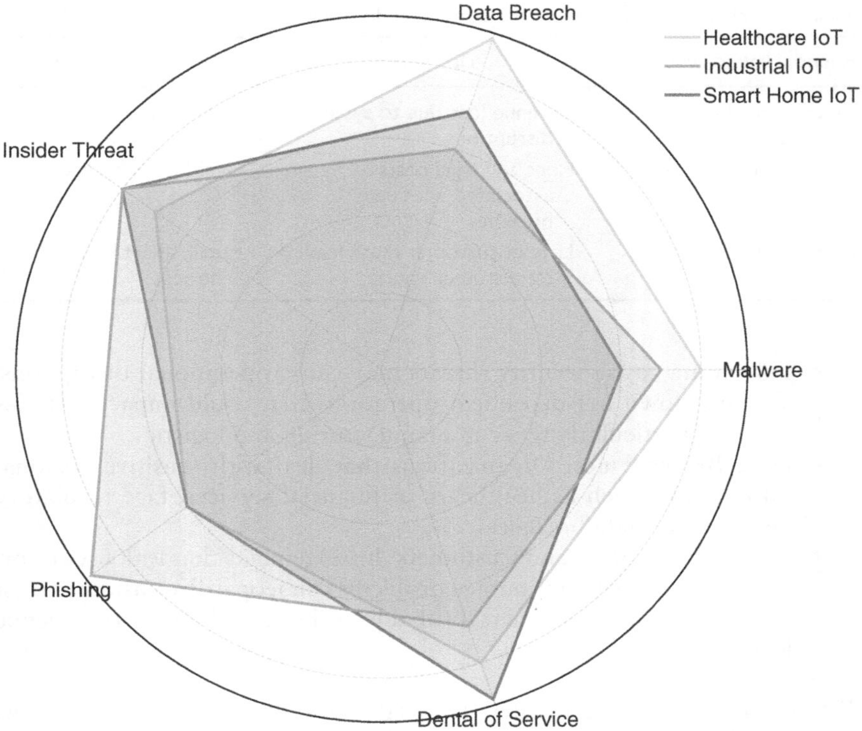

Figure 3.15 Data privacy impact from security threats.

- **Loss of data confidentiality**: Inadequate data encryption allows an attacker to intercept and view sensitive data in transit, breaching data confidentiality and increasing the chances of data manipulation.
- **Integrity of data compromise**: Attacks that modify or corrupt data compromise reliability may cause false readings or lead to wrong decision-making with potentially devastating consequences, especially in medical and industrial applications.

Figure 3.15 could display the extent of the impact on data privacy, showing which applications (e.g., healthcare, industrial, smart home) are most affected by each type of data privacy threat.

3.7.4 Financial impact

The financial fallout of IoT security threats is vast, with costs accruing from downtime, device repair, and legal liabilities encompassing loss of business. Table 3.18 provides the financial impact of security threats in IoT environments.

Table 3.18 Financial impact of security threats in IoT environments

Financial impact type	*Description*	*Affected industries*
Operational Downtime Costs	Revenue loss due to system disruptions	Manufacturing, logistics
Data Breach Penalties	Fines and legal costs associated with data breaches	Healthcare, finance
Reputational Damage	Loss of customer trust, leading to customer churn	Retail, smart home devices

- **Downtime costs**: Security threats may cause operational disruptions, shutting down IoT-dependent operations that could impact business revenue, particularly those in manufacturing and logistics.
- **Data Breach Fines**: Organizations that deal with sensitive personal information, such as healthcare or financial services, face regulatory penalties for data breaches.
- **Loss of Reputation and Customers**: Infringements that leak user information or cause major outage problems can jeopardize customer trust and result in lost customers, probably at the cost of long-term revenue losses.

Figure 3.16 represents estimated financial losses for operational downtime, data breach penalties, and reputational damage, giving a comparative view of how different threats translate to economic losses.

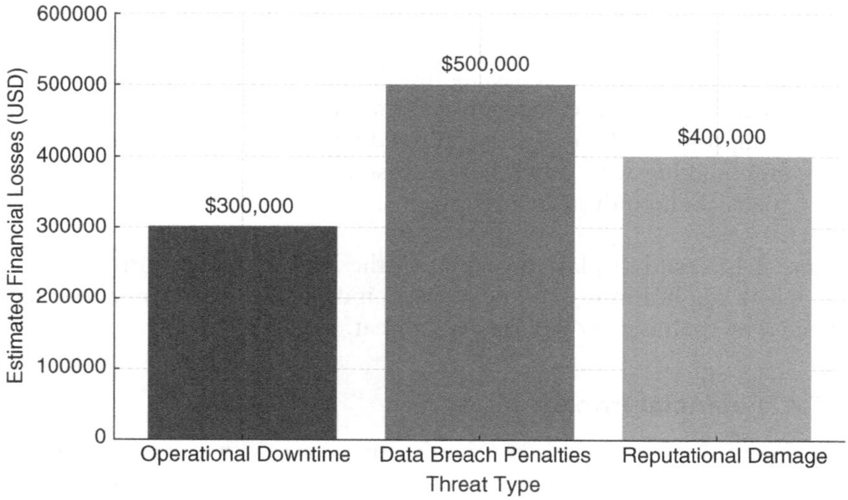

Figure 3.16 Financial costs associated with IoT security threats.

Table 3.19 Impact on user trust and IoT adoption due to security threats

Impact type	*Description*	*Affected sectors*
Decreased Consumer Confidence	Consumers avoid adopting IoT due to security concerns	Smart home, wearable tech
Business Reluctance	Companies delay adoption due to risk aversion	Industrial IoT, supply chain automation
Increased Demand for Security	Consumers prioritize secure devices, driving market changes	All IoT sectors

3.7.5 Impact on user trust and adoption

Security threats in IoT environments significantly impact user trust, impacting the adoption rate of IoT technologies. Concerns about data privacy, device reliability, or the potential for abuse may impede the willingness of individuals or organizations to adopt IoT systems. Table 3.19 shows the impact of security threats on user trust and IoT adoption.

- **Decreased consumer confidence**: Security breaches, especially of a data-privacy nature, have caused consumers to be mistrustful and hesitant to use IoT products in personal applications, such as smart homes.
- **Business reluctance**: The security risks and possible costs associated with breaches could discourage businesses from adopting IoT solutions, slowing innovation and realizing IoT benefits.
- **Increased demand for security features**: Rising security concerns lead consumers to demand more robust security features in IoT devices, which can increase production costs and foster more secure technology development.

Figure 3.17 shows how high-profile IoT breaches have affected consumer confidence over time, with data points corresponding to significant incidents and their effects on adoption rates.

3.7.6 Summary of security threat impacts across domains

Security threats in IoT environments have multifaceted impacts beyond simple device malfunction or data loss. As this section outlines, the consequences of IoT security threats ripple through device functionality, network performance, data privacy, financial stability, and user trust. Each impact type influences different sectors, affecting the adoption and evolution of IoT technology. Table 3.20 provides an overview of the impact of IoT security threats across key domains.

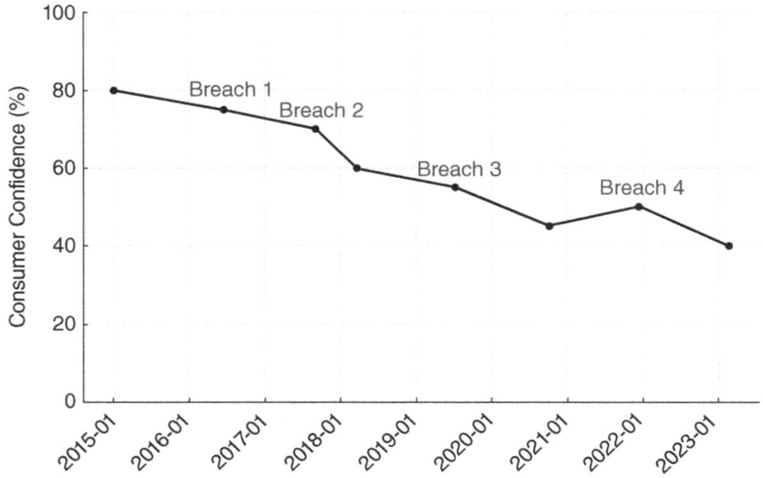

Figure 3.17 Change in consumer confidence over time due to IoT security threats.

Table 3.20 Overview of IoT security threat impacts across key domains

Impact domain	*Primary consequence*	*Key threat types involved*
Device Functionality	Malfunction, increased maintenance	Malware, ransomware, firmware manipulation
Network Performance	Network congestion, increased latency	DoS/DDoS, packet sniffing
Data Privacy	Loss of sensitive information, integrity compromise	Unauthorized access, data breaches
Financial Stability	Revenue loss, penalties, reputational damage	Operational downtime, data breaches
User Trust and Adoption	Reduced consumer confidence, slowed adoption	Privacy threats, lack of secure devices

Figure 3.18 provides a comprehensive view of the impact across each domain, showing which domains are most affected and quantifying the extent of each impact type.

Security threats in IoT environments have far-reaching impacts on device performance, network efficiency, data privacy, financial stability, and user trust. Effective security strategies must consider these impacts holistically since the increasing interrelation of IoT systems makes vulnerabilities in one area prone to propagating quickly to others. This chapter will expound in detail on these impacts so that IoT developers, organizations, and users can prioritize security features that would help ensure a safer ecosystem for IoT.

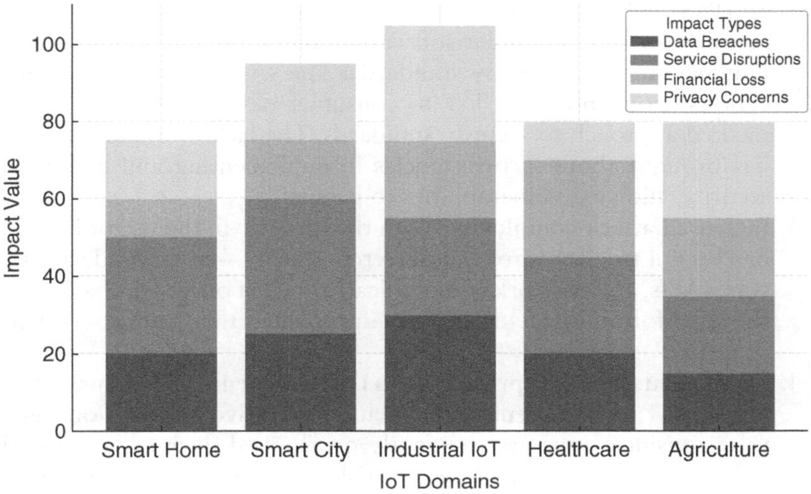

Figure 3.18 Comprehensive impact of security threats on IoT domains.

3.8 DISCUSSION AND RECOMMENDATIONS

The IoET has revolutionized industries and personal ambient environments with unprecedented automation, data sharing, and device interconnectivity. However, the proliferation of IoT has given way to several complex security challenges that threaten device functionality, data privacy, network performance, financial stability, and user trust. Only the inherent vulnerabilities of IoT devices and networks demand a more excellent picture of these threats and comprehensive mitigation strategies implemented. This section will henceforth discuss the issues of security challenges in IoT environments, analyze the key findings, and propose some practical recommendations for stakeholders to enhance security in IoT systems.

3.8.1 Discussion on the current state of IoT security

The landscape of IoT security is all about filling in the significant gaps in the current defense mechanisms. This is partly due to the unique architecture of IoT networks and the limitations of each device. The nature of IoT devices is usually constrained by processing power, minimal storage, and restricted operating systems; hence, traditional security solutions may not be applicable.

1. **Proliferation of devices with limited security capabilities:** Most IoT devices emphasize cost-efficiency and functionality over robust security features. This limitation, in turn, makes them easy targets for some

attackers who take advantage of minimal encryption, weak access controls, or lack of regular software updates.
2. **Fragmentation in security standards**: The wide variety of IoT applications, from Industrial IoT to consumer devices, has led to a fragmented approach to security standards. The lack of unified standards has brought about inconsistencies in implementing and maintaining security, allowing vulnerabilities to persist.
3. **Increased attack complexity**: With the increase in the use of IoT, there has been a trend toward multilayered attacks—targeting device firmware, APIs, and network communications all at once. This has exposed the need for multifaceted security approaches that protect every level of the IoT ecosystem.
4. **Insufficient focus on privacy in IoT deployments**: The constant data flow in IoT environments often includes sensitive information, yet privacy considerations are frequently overlooked during system design and implementation. IoT devices risk exposing personal and operational data to unauthorized parties without stringent data privacy measures.

3.8.2 Implications for key stakeholders

The IoT security threats span several classes of stakeholder groups, with differing effects felt by each due to vulnerabilities and risks associated with IoT systems.

1. **Consumers and end-users**: Users of IoT devices in smart homes, wearables, and personal applications are directly affected by breaches that expose their data or break the device's functionality. User privacy threats and diminishing user trust in IoT technologies could slow down the adoption rates of these technologies.
2. **Businesses and industries**: Organizations utilizing IoT systems in manufacturing, supply chains, and service delivery face operational disruptions and financial losses due to security threats. For example, a compromised industrial IoT system could halt production, leading to significant revenue losses and safety hazards.
3. **Government and regulatory bodies**: All regulators shall develop policies and guidelines for the security and privacy of IoT deployments. Without proper regulations, IoT devices will continue to mushroom without adequate security standards, posing a threat to public safety and national security.
4. **Manufacturers and developers**: The manufacturers and developers of IoT devices play a crucial role in integrating security features at the design level. The need for secure-by-design approaches is compelling since it's rather challenging to retrofit devices with strong security once deployed.

Table 3.21 Recommended device-level security enhancements

Device-level recommendation	*Description*	*Expected outcome*
Strong Authentication and Access Control	Enforces user identity verification	Reduces unauthorized access
Firmware and Software Updates	Regular updates to patch vulnerabilities	Protects against known exploits
Secure Hardware Components	Integrates hardware-based security measures	Enhances resilience against physical attacks

3.8.3 Recommendations for IoT security improvement

To tackle security problems in IoT environments, regulatory standards, technology innovation, and multilateral collaboration are needed. The following recommendations delineate a multitiered approach to establishing better IoT security.

3.8.3.1 Device-level security enhancements

This will require bolstering security at the device level because vulnerabilities at this level can be used to infiltrate the broader IoT networks. Table 3.21 provides recommended device-level security enhancements.

- Implement strong authentication and access control: To prevent unauthorized access, the manufacturer should implement multifactor authentication and unique access credentials for every IoT device. Secure boot processes may also be introduced to protect the devices from unauthorized firmware manipulation.
- Regular firmware and software updates: IoT devices must support remote firmware and software updates to patch security vulnerabilities continuously. Automated update mechanisms can help ensure devices remain protected against emerging threats.
- Secure hardware components: Integrate hardware-based security, including Trusted Platform Modules (TPMs), into IoT devices to provide secure storage for sensitive information, enhancing device integrity even during network compromise.

3.8.3.2 Network-level security measures

IoT network security will help prevent unauthorized access and reduce threats to data communication between devices.

- End-to-end Encryption: Strong encryption protocols, like AES-256, must be implemented to ensure that data in transit between IoT devices is unreadable to unauthorized users and to reduce the risk of man-in-the-middle attacks.

Table 3.22 Recommended network-level security measures

Network-level recommendation	*Description*	*Expected outcome*
End-to-End Encryption	Protects data in transit from interception	Secures data confidentiality
Network Segmentation	Limits the spread of attacks across the network	Contains threats to isolated segments
Intrusion Detection and Prevention	Monitors and prevents unauthorized network access	Enhances threat detection and response

- Network Segmentation: Isolating sub-networks within IoT environments will hinder attackers from moving laterally across the network. Isolating critical devices from unsecured devices will also help contain potential security breaches.
- Intrusion Detection and Prevention Systems (IDPS): IDPS can be deployed to monitor network traffic for suspicious activity. This will allow the early detection of threats and response to potential attacks.

Anomaly-based IDPS offers excellent detection capabilities, specifically for IoT networks, as it flags unusual patterns. Table 3.22 shows the recommended network-level security measures.

3.8.3.3 Data privacy and compliance

Data privacy protection in IoT ecosystems is critical, given the large volumes of sensitive information generated by these devices.

- Data Minimization and Anonymization: IoT devices are designed to collect only the necessary data, and data shall be anonymized wherever possible before transmission to reduce privacy risks.
- Data Encryption at Rest: Sensitive data on devices needs to be encrypted to prevent unauthorized access in case a device is compromised or stolen.
- Adherence to Privacy Regulations: IoT manufacturers and organizations should respect privacy regulations such as GDPR and HIPAA to conduct reasonable data handling practices. Compliance with these standards can also help mitigate potential legal liabilities. Table 3.23 provides recommended data privacy and compliance practices.

Table 3.23 Recommended data privacy and compliance practices

Data privacy recommendation	*Description*	*Expected outcome*
Data Minimization and Anonymization	Collect only essential data and anonymize	Reduces risk of sensitive data exposure
Encryption of Data at Rest	Protects stored data with encryption	Enhances data confidentiality
Compliance with Privacy Regulations	Ensures adherence to data protection laws	Mitigates legal risks and improves trust

3.8.3.4 Standardization and regulatory compliance

Uniform regulations and compliance requirements are essential to build a coherent security framework for IoT environments.

- **Development of IoT security standards**: Governments and regulatory bodies should establish minimum security requirements for IoT devices to ensure baseline protection across all devices.
- **Mandatory security certification**: To create a benchmark for secure devices, IoT devices may be required to undergo security certification and comply with set regulations, such as the IoT Cybersecurity Improvement Act.
- **Public awareness campaigns on IoT security**: A public awareness campaign can significantly help increase consumers' and businesses' awareness of IoT security risks, thus encouraging informed purchasing decisions toward more secure devices. Table 3.24 provides standardization and regulatory recommendations for IoT security.

Table 3.24 Standardization and regulatory recommendations for IoT security

Regulatory recommendation	*Description*	*Expected outcome*
Development of Security Standards	Establishes baseline security requirements	Ensures consistency across devices
Mandatory Security Certifications	Requires certification for device security	Drives security-focused manufacturing
Public Awareness Campaigns	Educates users on security best practices	Increases demand for secure devices

IoT smart environments have great potential but are currently handicapped by severe security vulnerabilities that compromise the device's reliability, data integrity, and users' privacy. Integration of strong security measures at device, network, and data privacy levels and support from the regulatory bodies through which standardized security practices are implemented can go a long way in improving the security posture of IoT environments. The key to achieving a secure and trusted IoT ecosystem will be collaborative efforts from manufacturers, businesses, governments, and end-users. Following these recommendations not only secures individual devices but also strengthens the security of the entire IoT environment, which should foster innovation and adoption.

3.9 CONCLUSION

The rapid development of the Internet of Things technology has brought revolutionary change across the sectors of healthcare, manufacturing, and smart cities to home automation. IoT devices provide effortless connectivity and data-driven decision-making, enabling systems to interact with and respond to real-world conditions in real time; however, this great potential also comes with substantial security challenges. This chapter has looked at the significant security threats that face IoT smart environments, evaluated their impact on device functionality, network integrity, data privacy, and user trust, and proposed a framework of recommendations to enhance IoT security at the device, network, and regulatory levels.

3.9.1 Summary of key findings

There have been critical security threats identified and discussed within IoT environments in this research chapter; these are summarized below:

1. **Complex and multilayered threat landscape:** The IoT smart environment is exposed to several security threats, such as unauthorized access, malware attacks, denial-of-service attacks, data breaches, man-in-the-middle attacks, etc. Such threats are usually exploited through the unique features that characterize IoT devices, such as limited processing power, weak encryption, and infrequent software updates, to compromise system security and privacy.
2. **Important consequences for device functionality and network performance:** Security threats will likely disrupt device functionality, causing malfunction, increased maintenance needs, and operational downtime. In addition, network performance will be affected, with bandwidth exhaustion leading to latency issues and congestion in the network that reduce the efficiency and reliability of IoT systems.

3. **Risk to privacy and data integrity**: IoT environments handle huge volumes of sensitive data, making data privacy and integrity the most critical concerns. Security breaches that expose or compromise such data may result in privacy violations, data manipulation, and misuse of sensitive information about individuals and organizations.
4. **Financial implication and erosion of user trust**: IoT security threats have significant financial implications due to costs from operational disruptions, regulatory penalties, and reputational damage. Their persistence further erodes user trust, an essential condition for IoT technologies' broader adoption and success.

3.9.2 Importance of a multifaceted security approach

As this chapter analysis indicates, the overall security of IoT environments calls for a multifaceted security approach. This is because security at the device level is not enough to address threats that find their way into IoT systems via network vulnerabilities and weak authentication and data protection measures. Hence, a holistic approach toward IoT security must address each layer of the IoT ecosystem, ensuring protection from the device and network level to data privacy and regulatory compliance. This could enhance the overall security posture and make the IoT systems more resilient to evolving cyberthreats

3.9.3 Recommendations to stakeholders

Securing IoT smart environments is a shared responsibility, requiring coordinated efforts from multiple stakeholders:

- **Manufacturers and developers**: Integrating security at the design stage, prioritizing secure firmware updates, and adhering to industry best practices can significantly reduce vulnerabilities in IoT devices.
- **Businesses and organizations**: Entities that deploy IoT technologies shall be committed to the highest order of security—through network-level defenses, routine vulnerability assessments, and adherence to data privacy regulations—to safeguard user information.
- **Regulators and policymakers**: The government and regulatory bodies have a vital role in setting standards and enforcing IoT security, ensuring all devices have a minimum level of security before being placed on the market. Moreover, promoting security certifications of IoT devices can drive manufacturers to adopt principles of security by design.
- **End-users and consumers**: Since end-users can make informed choices and prioritize secure IoT devices, this will eventually incentivize manufacturers to integrate better security into their products.

3.9.4 Future directions and research needs

As the technology of IoT will continue to advance, so will the threats that target such environments. Research into secure IoT architectures, lightweight encryption protocols, and security solutions driven by artificial intelligence will be fundamental to keeping pace with new and sophisticated attack methods. Standardizing security practices across IoT sectors will also help create a consistent protection framework, making it easier for organizations to implement adequate security measures.

Future research should focus on developing advanced, scalable security solutions tailored to the resource constraints of IoT devices. It will also be necessary to further develop threat intelligence platforms specific to IoT, along with real-time monitoring systems that will help detect and respond to threats more effectively. Further, research into user-centric privacy frameworks can ensure that data privacy is given priority when IoT adoption grows.

In the final analysis, IoT smart environments bear transformative potential. Still, they are poised against a myriad of security threats that could put at risk the functionality of devices, data integrity, and user trust. Those threats will demand a collaborative approach from manufacturers, organizations, regulators, and consumers in building secure IoT ecosystems. With strong security measures in place, following the best practices of the industry, and with a regulatory framework that encourages security and privacy, the IoT industry can prevent these risks and make all the potential of IoT technology available, ensuring the creation of a secure, reliable, and user-centered IoT future. With proactive and sustained efforts, one can protect IoT environments and create a safer and more connected world.

REFERENCES

[1] Ramadan R. A., Haidar Sharifa M., and Salem M.S., "SIoT: Secure IoT Framework for Smart Environments," in *Emerging Technologies in Computing*, vol. 332, M. H. Miraz, P. S. Excell, A. Ware, S. Soomro, and M. Ali, Eds., in Lecture Notes of the Institute for Computer Sciences, Social Informatics and Telecommunications Engineering. Cham: Springer International Publishing, 2020, pp. 51–61. doi: 10.1007/978-3-030-60036-5_4

[2] Karie N. M., Sahri N. M., Yang W., Valli C., and Kebande V. R., "A review of security standards and frameworks for IoT-based smart environments," *IEEE Access*, vol. 9, pp. 121975–121995, 2021.

[3] Fazio M., Celesti A., Puliafito A., and Villari M., "Big data storage in the cloud for smart environment monitoring," *Procedia Comput. Sci.*, vol. 52, pp. 500–506, 2015.

[4] Elrawy M. F., Awad A. I., and Hamed H. F. A., "Intrusion detection systems for IoT-based smart environments: a survey," *J. Cloud Comput.*, vol. 7, no. 1, p. 21, Dec. 2018, doi: 10.1186/s13677-018-0123-6

[5] Ahmed E., Yaqoob I., Gani A., Imran M., and Guizani M., "Internet-of-things-based smart environments: state of the art, taxonomy, and open research challenges," *IEEE Wirel. Commun.*, vol. 23, no. 5, pp. 10–16, 2016.
[6] Briante O. et al., "A Social and Pervasive IoT Platform for Developing Smart Environments," in *The Internet of Things for Smart Urban Ecosystems*, F. Cicirelli, A. Guerrieri, C. Mastroianni, G. Spezzano, and A. Vinci, Eds., in Internet of Things. Cham: Springer International Publishing, 2019, pp. 1–23. doi: 10.1007/978-3-319-96550-5_1
[7] Kubitza T., "Apps for Environments: Running Interoperable Apps in Smart Environments with the meSchup IoT Platform," in *Interoperability and Open-Source Solutions for the Internet of Things*, vol. 10218, Podnar Žarko I., Broering A., Soursos S., and Serrano M., Eds., in Lecture Notes in Computer Science. Cham: Springer International Publishing, 2017, pp. 158–172. doi: 10.1007/978-3-319-56877-5_10
[8] Miloslavskaya N. and Tolstoy A., "Internet of Things: information security challenges and solutions," *Clust. Comput.*, vol. 22, pp. 103–119, 2019.
[9] Hajjaji Y., Boulila W., Farah I. R., Romdhani I., and Hussain A., "Big data and IoT-based applications in smart environments: A systematic review," *Comput. Sci. Rev.*, vol. 39, p. 100318, 2021.
[10] Verma A., Khanna A., Agrawal A., Darwish A., and Hassanien A. E., "Security and Privacy in Smart City Applications and Services: Opportunities and Challenges," in *Cybersecurity and Secure Information Systems*, A. E. Hassanien and M. Elhoseny, Eds., in Advanced Sciences and Technologies for Security Applications. Cham: Springer International Publishing, 2019, pp. 1–15. doi: 10.1007/978-3-030-16837-7_1
[11] Ystgaard K. F. and De Moor K., "Envisioning the future: a multi-disciplinary approach to human-centered intelligent environments," *Qual. User Exp.*, vol. 8, no. 1, p. 11, Dec. 2023, doi: 10.1007/s41233-023-00064-5
[12] Caviglione L., Lalande J.-F., Mazurczyk W., and Wendzel S., "Analysis of Human Awareness of Security and Privacy Threats in Smart Environments," in *Human Aspects of Information Security, Privacy, and Trust*, vol. 9190, T. Tryfonas and I. Askoxylakis, Eds., in Lecture Notes in Computer Science. Cham: Springer International Publishing, 2015, pp. 165–177. doi: 10.1007/978-3-319-20376-8_15
[13] Faruqui N. et al., "Cloud IaaS optimization using machine vision at the IoT edge and the grid sensing algorithm," *Sensors*, vol. 24, no. 21, p. 6895, 2024.
[14] Vinod Kumar T. M., Ed., Smart Environment for Smart Cities. in *Advances in 21st Century Human Settlements*. Singapore: Springer Singapore, 2020. doi: 10.1007/978-981-13-6822-6
[15] Al Razib M., Javeed D., Khan M. T., Alkanhel R., and Muthanna M. S. A., "Cyber threats detection in smart environments using SDN-enabled DNN-LSTM hybrid framework," *IEEE Access*, vol. 10, pp. 53015–53026, 2022.
[16] Kimani K., Oduol V., and Langat K., "Cyber security challenges for IoT-based smart grid networks," *Int. J. Crit. Infrastruct. Prot.*, vol. 25, pp. 36–49, 2019.
[17] Kabalci Y., Kabalci E., Padmanaban S., Holm-Nielsen J. B., and Blaabjerg F., "Internet of things applications as energy internet in smart grids and smart environments," *Electronics*, vol. 8, no. 9, p. 972, 2019.
[18] Lin H. and Bergmann N. W., "IoT privacy and security challenges for smart home environments," *Information*, vol. 7, no. 3, p. 44, 2016.

[19] Ahmad W., Vashist A., Sinha N., Prasad M., Shrivastava V., and Muzamal J. H., "CrowdFaceYOLO: Advancing Real-Time Face Detection in High-Density Crowded Area," in *2024 Artificial Intelligence for Business (AIxB)*, IEEE, 2024, pp. 92–93. Accessed: Dec. 22, 2024. [Online]. Available: https://ieeexplore.ieee.org/abstract/document/10771193/?casa_token=yTjllc8uYbkAAAAA:ozCfBmriXNmpWB9_iQP1f6UX4HY2A7vvSsLqm-H8TdDEPXaIEZEEIhfrtGOGg_2za1GWFyEWYFKNXg

[20] Hazman C., Guezzaz A., Benkirane S., and Azrour M., "Toward an intrusion detection model for IoT-based smart environments," *Multimed. Tools Appl.*, vol. 83, no. 22, pp. 62159–62180, 2024.

[21] Hussain A., Aslam A., Tripura S., Dhanawat V., and Shinde V., "Weather forecasting using machine learning techniques: Rainfall and temperature analysis," *J. Adv. Inf. Technol.*, vol. 15, no. 12, 2024, Accessed: Dec. 22, 2024. [Online]. Available: https://www.preprints.org/frontend/manuscript/6cd41fc835f5c2834d9a37dacc41c4df/download_pub

[22] Srivastava M. and Kumar R., "Smart Environmental Monitoring Based on IoT: Architecture, Issues, and Challenges," in *Advances in Computational Intelligence and Communication Technology*, vol. 1086, X.-Z. Gao, S. Tiwari, M. C. Trivedi, and K. K. Mishra, Eds., in Advances in Intelligent Systems and Computing. Singapore: Springer Singapore, 2021, pp. 349–358. doi: 10.1007/978-981-15-1275-9_28

[23] Mehmood H., Khalid A., Kostakos P., Gilman E., and Pirttikangas S., "A novel Edge architecture and solution for detecting concept drift in smart environments," *Future Gener. Comput. Syst.*, vol. 150, pp. 127–143, 2024.

[24] Li D., Luo Z., and Cao B., "Blockchain-based federated learning methodologies in smart environments," *Clust. Comput.*, vol. 25, no. 4, pp. 2585–2599, Aug. 2022, doi: 10.1007/s10586-021-03424-y

[25] Fadi O., Karim Z., and Mohammed B., "A survey on blockchain and artificial intelligence technologies for enhancing security and privacy in smart environments," *IEEE Access*, vol. 10, pp. 93168–93186, 2022.

[26] Gunduz M. Z. and Das R., "Cyber-security on smart grid: Threats and potential solutions," *Comput. Netw.*, vol. 169, p. 107094, 2020.

[27] Premkumar M., Ashokkumar S. R., Mohanbabu G., Jeevanantham V., and Jayakumar S., "Security behavior analysis in web of things smart environments using deep belief networks," *Int. J. Intell. Netw.*, vol. 3, pp. 181–187, 2022.

[28] Mohy-eddine M., Guezzaz A., Benkirane S., and Azrour M., "IoT-Enabled Smart Agriculture: Security Issues and Applications," in *Artificial Intelligence and Smart Environment*, vol. 635, Y. Farhaoui, A. Rocha, Z. Brahmia, and B. Bhushab, Eds., in Lecture Notes in Networks and Systems. Cham: Springer International Publishing, 2023, pp. 566–571. doi: 10.1007/978-3-031-26254-8_82

[29] Perera C., Jayaraman P. P., Zaslavsky A., Christen P., and Georgakopoulos D., "Context-Aware Dynamic Discovery and Configuration of 'Things' in Smart Environments," in *Big Data and Internet of Things: A Roadmap for Smart Environments*, vol. 546, N. Bessis and C. Dobre, Eds., in Studies in Computational Intelligence. Cham: Springer International Publishing, 2014, pp. 215–241. doi: 10.1007/978-3-319-05029-4_9

[30] Shao Y., Lessio N., and Morris A., "Iot avatars: Mixed reality hybrid objects for core ambient intelligent environments," *Procedia Comput. Sci.*, vol. 155, pp. 433–440, 2019.

[31] Shahrestani S., "The IoT and Smart Environments: An Overview," in *Internet of Things and Smart Environments*, Cham: Springer International Publishing, 2017, pp. 57–73. doi: 10.1007/978-3-319-60164-9_4
[32] Mitra S., Chaulya S. K., Kumar D., and Soni A., "An Approach for Implementation of IoT Enables Smart Environmental Monitoring and Strata Monitoring System for Underground Coal Mines," in *Proceedings of the 10th Asian Mining Congress 2023*, A. Sinha, B. C. Sarkar, and P. K. Mandal, Eds., in Springer Proceedings in Earth and Environmental Sciences. Cham: Springer Nature Switzerland, 2023, pp. 165–179. doi: 10.1007/978-3-031-46966-4_14
[33] Ismagilova E., Hughes L., Rana N. P., and Dwivedi Y. K., "Security, privacy and risks within smart cities: Literature review and development of a smart city interaction framework," *Inf. Syst. Front.*, vol. 24, no. 2, pp. 393–414, Apr. 2022, doi: 10.1007/s10796-020-10044-1

Chapter 4

Governance frameworks for artificial intelligence of things (AIoT) security

Wasswa Shafik

4.1 INTRODUCTION

AIoT refers to the integration of artificial intelligence (AI) technologies with the Internet of Things (IoT), making it possible for AI agents to monitor, learn, control, and communicate with things. Due to the paradigm shift brought by greater intelligence, AIoT is taking center stage. AIoT is also driving new business models and relationships between stakeholders [1]. However, the rapid technology convergence during AIoT systems design brings new threats and vulnerabilities that are evolving. The main cause of these threats and vulnerabilities in AIoT is the lack of security, taking into account the intersection of artificial intelligence and IoT. Threats of AIoT can include smart botnets, AI-based malware, adversarial attacks, and intelligent AI [2]. Undoubtedly, traditional perimeters and security mechanisms have not provided the required security level for AIoT. Consequently, the accelerated AIoT innovation networks from industry and academia aim to accelerate the establishment of a security framework that could support the principle of data protection by design and by default. It is crucial due to the importance and rapid technology convergence nature of AIoT, coupled with the spectrum of possible applications. This report delves into both the necessity and implications of AIoT security governance [3].

In conclusion, facilitating the adoption of AIoT systems critically depends on addressing the increasingly sophisticated threats and regulatory compliance that are possible through robust AIoT security. Effective governance of AIoT can deliver several tangible benefits, including a significant increase in the robustness of cybersecurity and increased trust in AIoT from both the public and private sectors. An organization is as secure as its cybersecurity team. At the core of every cybersecurity team, there is a specific approach to managing AIoT security, given its particular complexity on a framework basis [4]. Fostering trust in technology is a task that goes beyond good technical mechanisms and security measures and extends to adopting a set of principles and practices at all levels of the system's life that can demonstrate with minimal doubt to all parties involved that the technology is

DOI: 10.1201/9781003606307-4

trustworthy [5]. That said, there is no doubt that AIoT poses greater risks than either AI or IoT technologies considered separately, and the absence of adequate tools to mitigate this heightened level of risk is already leading to a lack of trust trend, as reflected in painful withdrawals of technological investment. To address these issues, executives need a governance framework that balances risk, security policies, and best practices that guide the supply chain players to engage with the AIoT environment as securely as possible, address and solve risks as they appear, and act in unison to engage in this trust-evoking approach [1, 2].

4.1.1 Overview of AIoT and its security challenges

The term "Artificial Intelligence of Things (AIoT)" refers to the combination of Artificial Intelligence (AI) techniques and Internet of Things (IoT) technologies to enable smart or intelligent autonomous systems to interact with each other and with humans. These AIoT systems operate with very limited human intervention, meaning that it is imperative that they can manage their security [6]. Furthermore, because AIoT systems typically operate in dynamic, open environments, they may need to be reconfigurable and even learn. The proliferation of AIoT systems is constrained by fundamental security challenges that AIoT technology developers and adopters need to address. When exchanged, the data can be encrypted, signed, and timestamped so that it is tamperproof and reflects some minimum level of data confidence and integrity [7]. However, if the underlying data processing and AI models are compromised, then it is not possible to ensure the robustness or safety of the generated outcomes.

AIoT systems could thus be vulnerable to threats such as data tampering, unauthorized access and use, inferring unauthorized outputs, data breaches, identifying patterns, revealing sensitive data, facial spoofing, privacy violations, model inversion, adversarial learning or poisoning, app penetration, infrastructure penetration, distributed denial of service, and others [8]. Such vulnerabilities and the potential risks and impacts associated with them imply a multifaceted effective risk management requirement, including legal compliance, privacy risk management, cybersecurity, and physical safety. While traditional cybersecurity measures and privacy-enhancing technologies may offer some protection against these threats, they will not cover the full range of such risks and may undermine AIoT intelligence through side-channel defenses [9]. The connected and autonomous nature of AIoT also potentially creates new cascades and amplifications of risks that need to be anticipated and managed.

Finally, threat actors are also highly dynamic and continuously evolve, which could require the same for security measures and controls. Given these characteristics, novel governance mechanisms and models need to be considered so that AIoT systems can be secure and compliant in practice.

The chapter describes the outline of many related works, governance required, standards, and protocols. Additionally, AIoT system security risks are described for a complete method to be designed to address the found requirements [10, 11]. This work should attract security researchers, diverse industrial stakeholders, government entities, and many more to secure AIoT systems effectively. AIoT governance is as important as securing AIoT systems to operate lawfully and for the safety of humans, intellectual property, and the involved devices, data, and infrastructure [11, 12]. Ensuring the trustworthiness and safety of AIoT systems will definitely play a significant role in the wide acceptance of AIoT and the growth of the AIoT economy.

4.1.2 Importance of governance frameworks in AIoT security

The governance framework addresses the how of AIoT security governance. It is handy in establishing ways and means of how security management can ensure the security of the underlying AIoT. It is essentially a structured way of defining protocols, policies, procedures, and principles for all stakeholders and can be applied consistently to demonstrate the organization's commitment to a better AIoT world. Governance frameworks establish clear lines of accountability and responsibility to boost stakeholders' and users' confidence [13]. The policies and laws also define liability and responsibility, which refer to governance in a model: Governance defines what is to be done and who is held responsible, which directs management to determine how to act and set goals and objectives, and ensures that "what needs to be done is being done," and puts in place appropriate means of coordination. So, for an organization driven by governance, accountability, and responsibility lie directly with the governing body to secure the freedom of users and their interest in AIoT security management [14].

The governance frameworks need to have basic elements to programmatically lay out the role and responsibility of security stakeholders, which can also be used as legal evidence for providing directions to enterprises. The performance-driven policies and laws under AIoT systems and services encourage the creation of self-regulatory AIoT security management. Until a disaster happens, in a reactive system, it is very late to handle and manage such risks damaging entities [6]. However, proactive security management under the oversight of governance and a constant touch of AIoT technologies will create a conducive environment for innovation. It is potentially a driver of some funds to be invested by stakeholders toward AIoT innovation. It is a willingness to accept such an invention disrupting our lives or harming society due to the lack of security in AI-based services and products that are developing. If there exists no governance structure to put principles into application, society would resist, suspect, and fear this innovation [7, 8].

4.2 FUNDAMENTAL CONCEPTS OF AIoT SECURITY

AIoT involves the convergence of AI and IoT to expose AI functionalities to auxiliary systems. In this section, we briefly explain the fundamental technologies behind AIoT. AI includes algorithms or math-embodied instructions that articulate chatbots, natural language processing, recommendation systems, machine learning, vision computing, robotics, etc. AI is increasingly growing. AIoT improves the communications between AI components or assisted systems using IoT to enhance the implicit collaboration among heavily AI-dependent systems. Artificial Intelligence of Things will optimize new solutions for the general environment, including AI, cloud computing, IoT, and others [10, 11].

Managing a system's security needs a comprehensive understanding of the key properties that need to be secured. These potential security requirements generalize the system's goals and features. When discussing security for AIoT, we require a set of principles that ensure a logical approach to the governance of AIoT security. We consider the Confidentiality, Integrity, and Availability (CIA) principles as follows to derive some basic principles that could govern AIoT security due to four following reasons. The first aim of AIoT governance is to provide a logical explanation and should reflect the actual importance of Artificial Intelligence (AI) in an Internet of Things environment (IoT) [10, 11]. The second aim is to understand the relationship between the Internet of Things (IoT) basic principles and AI security operational principles. A third aim is to explain the basic concepts of AIoT governance to understand the developed framework or models clearly. Fourth, the CIA model has an impact on how the "governance" of AIoT can be defined specifically [14]. The applications and the systems having capabilities of both AI and IoT also showcase the negative aspects of AI along with its capabilities. By combining the abilities of AI, many IoT-based exploitations or attacks can happen in the systems. That is why to govern the AIoT, some basic principles are to be used. A few of these can be privacy, integrity, robustness, trustworthiness, dependability, recoverability, safety, etc. Confidentiality, integrity, and availability are key principles in managing IoT devices, ensuring secure data transmission, protection against unauthorized access, and continuous system functionality [6].

4.2.1 Artificial intelligence and internet of things technologies

The entities of interest in the field of Artificial Intelligence (AI) are agents that demonstrate their ability to perform tasks that require human intelligence, such as perception, cognition, or decision-making. AI systems are designed to learn, reason, and act to complement human decision-making or to replicate its function. The primary function of an AI system is to interpret vast amounts of data and understand relationships and interdependencies between substantial datasets using algorithms and rule-based decision-making [8]. With technological innovations, AI can handle tasks including diagnosing diseases, predicting natural disasters, personal assistants, etc.

The Internet of Things has basic components: sensors/actuators, controllers/gateways/RFIDs, and communication technologies for connectivity. These contribute to different functionalities of monitoring, controlling, and adapting things. The primary aim of IoT is to enable the connected world where most or all things with unique electronic identifiers have the capability of providing data about themselves. Both AI and the IoT are shaping a new area termed AIoT [9]. The integration of AI with IoT has the potential to bring valuable new insights based on data generated for better decision-making and real-time applications. In this context, AI empowers IoT by enabling smart and better context-based decision-making. While IoT involves extensive data processing and real-time decision-making, the inclusion of AI in the ecosystem can make the network "smarter" by enabling predictive and preventive network maintenance [11, 12].

4.2.2 Key security principles in AIoT

Confidentiality, integrity, and availability form the base security strategies for achieving an overall security and protective posture for all AIoT proceedings. Whenever AIoT frameworks, aims, or systems are developed to align with these three strategies, protection and security procedures will be guaranteed. To align with the C-I-A principles, various key operational and administrative security steps need to be applied [13, 14]. This could encompass, for instance, multistep verification mechanisms and methodologies, authoritative usage, regular patch distribution, procedural control setup, individual verification, and defense controls designed to:

- Guarantee the precision and compromise resistance of all users, roles, datasets, and other software and hardware components connected to protective technologies.
- Ensure the availability of reliable and consistent AIoT systems and services when required.
- Allow all measures to promote some areas of AIoT that rely on gossip, exchanging, speed, anonymity, and immutable smart data or distributed AI functions.

A risk governance approach based on communication and control with policies designed to evaluate, measure, and mitigate exposure and likelihood of the AIoT security risk to an acceptable level, as well as systematically range and degree, is fundamentally essential to AIoT in a connected AIoT ecosystem. Moreover, it is equally advantageous to sustain ethical concepts and regulations dedicated to protection and credibility in AIoT at an early conceptualization and establishment of secure AIoT platforms [15]. It takes possible security control initiatives on a blueprint for creating measures implicitly mapped. In conclusion, certain defensive security strategies should be carried out before an AIoT-connected item or procedure is designed to lower various technical and financial threats and obligations. AIoT technical and regulatory security measures ought to remain flexible, adaptable, and possibly strict with these defensive strategies as AI and IoT technologies are reasonably and swiftly evolving, assisting with patterns, tactics, strategies, and regulations to various sectors applying beneficial AIoT [8, 16].

4.3 SECURITY CHALLENGES IN AIoT

AIoT, a new technology, provides a holistic approach in a connected ecosystem such as smart homes, autonomous vehicles, and other applications. The increasing interplay between AI and IoT raises many security challenges due to the diverse and massive number of impacting sensors and devices. Security is the fundamental concern of AIoT from a privacy, confidentiality, integrity, and trust perspective [10]. Data privacy and data protection are of huge importance, especially in AIoT, where devices collect sensitive information. Hackers may exploit ultra-fast big data analytics and machine learning techniques in combination with the opportunities that arise from AIoT to track movement, gain access to normally closed buildings, and much more [7].

Attackers are continuously looking for opportunities to interleave into the workings of AIoT environments at different levels of the architecture. Integrating AI with IoT greatly increases the attack surface, including numerous unique vulnerabilities. AIoT devices and networks face attacks from various threats such as data breaches, unauthorized access, malware, or virus attacks interacting at different levels within the AIoT system, or sometimes in combination, leading to well-coordinated multilayered attacks such as social engineering at the human level of the ecosystem's initial layer and finally hacking in the IoT or AI [1, 2]. The interoperability caused by the integration of multiple devices in AIoT further leads to security challenges. Part of the major concern is security in AIoT, where researchers are still developing frameworks to address various security perspectives. It is required to develop more sophisticated security solutions in the area of AIoT. AIoT security must consistently protect assets, information, and data from human, application, or physical threats that would degrade privacy, confidentiality, and trust [3]. In an AIoT environment, we find very unique and complex security challenges with competing factors. Dynamic and ever-increasing threats add to the complexities that arise in part due to the ambitious and forward-thinking development of emerging technologies. There are huge challenges related to AIoT integration, analysis of sensor data, self-learning algorithms, and AIoT security. The framework provides proactive approach mechanisms as safeguards and is coined with the belief that "it's better to be ahead of the game." The strategies further add to the defensive posture of AIoT security [4].

4.3.1 Data privacy and protection

The collection of personal data is a critical part of the functioning of AIoT systems. This data may be sensitive, especially if collected through sensors embedded in wearables and implants, as is mostly common in AIoT environments. Indeed, the proliferation of Internet-enabled health-monitoring devices makes the privacy of biometric data a concern with the increasing misuse value in the underground market for cybercriminals. Data breaches in AIoT devices used for e-health can have far-reaching implications for the future mobilization of cybercrime due to the exclusive collection of data like health, personal habits, and individual body functionality [5, 17]. To ensure appropriate protection and fair use of personal data, national and international regulatory regimes have devised a plethora of legislative and policy frameworks. The imposition of extra obligations on organizations processing personal data gives a higher risk to the rights and freedoms of individuals that are inherent in the functioning of AIoT devices. Therefore, the development of trust in AIoT is inherently linked to good data privacy and protection controls by the systems' architects and organizations. Furthermore, every organization is under a strict obligation to handle personal data lawfully, cautiously, and transparently, with explicit limitations on the data collected and processed [18].

The technology primarily relies on the processing and gathering of decentralized and large volumes of data across multiple devices and operators. However, the effective application of data privacy principles is challenging due to the lack of a standardized stack inside the AIoT. The architecture and operation of devices and systems are often dependent on the developers' expertise, networking protocols, embedded systems, and software tools used alongside the organizational choices in terms of suppliers and partners [19]. Some of the strategies for data privacy risk reduction include the use of network ciphers and device encryption. The abstractions of device-to-device communications through intermediary AI relate to the security benefits of legacy web browsers in contemporary web communications. Identity anonymization with a decoy identity is another approach that is under speculation for the reduction of risks in e-health in particular. The consequences of the rising plethora of vulnerabilities range from the social aspect of personal behavior being leaked and exploited with the devices constantly worn and activated, ideally without being noticed [20]. This would render these devices criminal enablers, and the breach and use of such information in an ethical paradigm would become a global issue prior to being prioritized by any organization or individual. Hence, the consent of the users on what data is being collected and the purposes it fulfills is interactive. Without a clear, strong element of public consensus and awareness, the governing models can be seen as exclusionary. This risks ultimately perpetuating a trust circle of a minority at the very exclusive center and the majority on the periphery. In such a model, the reliance on the minority culminates in it being stopped by the majority due to the lack of trust. As seen in such applications before, the minority would be the active user base and, ultimately, the mature and developed AIoT [21, 22]. The majority is at immediate risk of extra uncertainty from considerations like job security and data consolidation by any one AIoT provider. This, in effect, would stop the very mass adoption these governing models would hope for.

4.3.2 Threats to AIoT systems

AIoT faces threats from several attack vectors that have been designed to target their AIoT characteristics. AIoT systems offer new and extended possibilities of how different attack vectors can be mounted, such as rogue devices or body area network hijacking. From the AIoT governance and risk management perspectives, the consequences of the attacks are most valuable in characterizing the threats. After such analyses, managers can order their risk management activities more precisely to counteract these threats [9]. Common attack vectors directly targeting AIoT devices are, for example, malicious software running built-in AI algorithms or software for facial recognition cases. Also, standard attacks like a Denial-of-Service attack directly targeting AIoT features like driving-oriented AI can be considered threats to AIoT. Common attack vectors for IoT in general, like DoS for flooding types of AIoT, are also compatible as threats to AIoT and can be considered directly. For example, if there is no control to that end, even a rogue device in one building can render the system vulnerable on a wide network [10, 11]. Because the AIoT ecosystem with its AIoT features can be considered a sum of IoT networks, the interconnectivity and weaknesses will also put vulnerable AIoT devices at risk.

4.3.3 Cyber-physical systems AI-enabled technologies

The importance of adopting a dynamic approach in governing AIoT can be further understood in the light of the cybersecurity threat framework: to attack the cyber-physical system integrity. This is insightful in that AI in cyber-physical systems connects cyberspace with the physical world, and the safety of the latter is networked to the security of AI algorithms and their applications. Cyber-physical systems AI-enabled technologies have vulnerabilities that adversaries can exploit to gain authorized access [12]. For example, the control signals can be manipulated and operated by a malicious actor, or the devices can deny access to authorized users using DoS and other similar attacks: data integrity is the only means of distinguishing between authorized and unauthorized access. As such, there is a need for governance within the distinct security framework specific to AIoT that covers communicating data errors, generating alerts based on data analysis errors, modifying the dataset by adding errors for eventual feedback into training data, and modifying device decision output by adding errors [12, 13]. Like IoT devices, AIoT devices like connected vehicles are vulnerable to attacks from adversarial AI. This can take the form of manipulating data or using vulnerabilities in retraining protocols and consequently adapting AI model behavior. In the appropriate context, connected vehicles represent impacted robotic infrastructure across many countries, and hence, this fragility in AI can represent a challenge to connected AIoT security. At the same time, the other side of AI requires credit, as there will be AI in cybersecurity software that can anticipate and automatically counteract such attacks, but there will be a gap between attacks and antivirus solutions for AIoT [14, 15]. Those challenges will require collaboration among interested stakeholders and conducting future research in this direction. In the following subsection, we analyze different AIoT security governance frameworks.

4.4 EXISTING AIoT SECURITY GOVERNANCE FRAMEWORKS

Governance frameworks for AIoT security can be organized into two categories: reactive and predictive. The reactive approach emphasizes that improvements can be based on incident reports and related measurements. In contrast, the predictive approach encompasses an aspect concerned with forecasting AIoT-related changes and their potential security implications. Governance frameworks can also be categorized as extensible, which assumes that the existing best practices for a selected area can be given and validated by studies in this field, and non-extensible, where systemic best practices from IT, IoT, and AI are provided to deliver global policies. Some general approaches for securing the AIoT environment and the role of AI in security are covered in theoretical overviews. From a technical standpoint, the remaining sections present the way to configure, train, and run one of the models proposed to predict AIoT security session outcomes, states, or record or generate samples that manage with the characteristic function of the applicable environment [23].

This section reviews existing AIoT security governance frameworks and offers a comparative analysis that gives insight into the potential and usefulness of different governance approaches. Furthermore, selected examples of AIoT governance frameworks, with a view on the approach, scalability, and application of these frameworks into practice, are presented. Two perspectives are highlighted in this survey: horizontal, where structures for data processing are considered, and vertical, where security is commonly considered a different subject for each level of a real deployment [20, 21]. The issues of data security and trust related to real deployment are considered in terms of enterprise, edge/fog network, sensors/actuators, and up to the modem.

4.4.1 Overview of current frameworks

The concept of "AIoT Governance" is developed to govern the AIoT in addressing AIoT security risks. Governance in a broad scope includes formal governance via regulations, legislation, and voluntary standards, as well as informal governance via codes of conduct, best practices, guidelines, principles, etc. Addressing AIoT-related challenges from the perspective of "governance" involves a range of stakeholders with expertise in respective areas such as IoT, AI, standardization bodies, industry consortia, and regulatory entities. Therefore, various governance frameworks are established by enterprises, organizations, or regulators to catalyze efforts that are needed to secure AIoT applications [19, 24]. It is notable that while telemetry frameworks specify the transmission mechanisms, metrics, and processing steps, not all refer to purpose definitions, nor do they monitor purposes via audit fees. In contrast, trust mechanism guidelines tend to integrate purpose monitoring into a wider framework as a part of establishing trust. We break down the objectives and initial applications of the representative AIoT governance frameworks in the next section. Each framework is developed to handle the security needs of specific AIoT systems: automotive, update, standard, and privacy-first, with a focus on data governance and trust management [25]. They aim to provide universally applicable security governance for various industries and jurisdictions and are developed in cooperation with industry and stakeholders.

4.4.2 Regulatory frameworks

Governments and their respective agencies are foundational in providing recommendations and regulations that require mandatory compliance. Notable regulations from governments with sections focused on AIoT security include data minimization, which incentivizes a proportional security posture, and guidelines on the cybersecurity of IoT devices and small businesses. These guidelines include various reports and documents developed by relevant governmental departments. As the guidelines have been translated into additional languages, these globally now reflect agreements over the soundness of the guidelines' recommendations [26]. Recommendations offer organizations a cutting-edge, regulation-aware posture for effective legal compliance regardless of the evolving climate of the legal landscape, influenced by the rapidly evolving AIoT landscape, despite the necessity of a gap analysis to maintain compliance ahead of enforcement actions [27]. Regulatory frameworks offer accountability for violations of ethical use for AIoT outside of auto-enforceable user consent contracts. Corporate and legal interest in defining the ethical use of technologies such as AIoT has been seen through compliance reports within various corporations and through intellectual property tribunals internationally applying norms to online surveillance and data processing technologies. In contrast, non-compliant ownership of personal data has fled incineration due to lingering public resentment in the form of data protection demands [28]. The benefits of holding corporations accountable to AIoT-driven value propositions, therefore, outweigh the costs of early adoption.

4.4.3 Industry standards

A wide range of organizations produce the creation and embedding of security, privacy, and trust TEDs. Industry standards facilitate the application of best practices and allow for the interoperation of devices developed in various organizations. Many groups and organizations create standards. The accompanying standards covering security and trust for the AIoT are equally diverse and varied. For example, a working group is developing a standard covering architectural aspects, including security, trust, and privacy for AIoT and AIoT supply chains. Standards for Situational Awareness developed by various organizations are examples of standards that, while not directly targeting security for the AIoT, are often referenced in proposals to establish AIoT security [29]. Organizations develop voluntary industry standards, and the users of these standards have no obligation to use them. The lack of regulatory implications means voluntary use and implementation are only as effective as the level of participation in the standard creation process allows. Conversely, hardware, software, or systems using voluntary industry standard work products will experience more credibility and have an easier path to qualification with regulatory agencies if they follow these standards [30].

4.4.4 Conflicting source standards

Implementing the guidance around ZPRs and ZPIs increases one's security posture. Alternatively, combining domain-specific regulatory and industry standards and creating security programs that encompass them provides an end user with a balanced approach to compliance and regulatory acceptance. Industry standards may be based on other standards documents; however, as they are professional opinion-based, in part, they may also contradict or overrule the source standards if a standard with a greater narrative is created from conflicting source standards [31]. An example of a standard that uses both regulatory and industry standards is a harmony between two sets of standards. Industry standards need some level of consensus, albeit voluntary and somewhat polarized. Leading economic sectors and road-mapped infrastructures must take part, or there will be no short-term path to standard creation. Industry stakeholders, businesses, academia, and government sectors need to work together to create consensus standards. Cross-industry alignment needs to expand and include big data, software companies, and other large stakeholder entities with an aligned roadmap [32, 33]. An example of the benefits of voluntary industry standards is the recent implementation to confirm its dedication to safety.

4.4.5 Comparison of different frameworks

This subsection aims to provide a comprehensive comparison and review of different AIoT security governance frameworks, standards, and certifications, taking into consideration regulatory compliance when applicable. The aim is to study their limitations and foundation in a richer context and gain, through this lens, a more detailed insight into effective technologies and their best practices in facilitating secure AIoT technologies and infrastructures. Gathering a clear picture of the technology's impact is important since the solutions offer various possibilities for policy implementation and governance effectiveness. We begin by discussing the different criteria used to compare and assess various technologies in the field of security governance [34]. It proceeds with examining the group of standards, frameworks, and certifications, with a conclusion discussing the importance of engagement and collaboration with stakeholders and the challenges faced when implementing harmonization of effective technologies. The frameworks and regulations that tech corporations can use to manage security must be assessed based on certain criteria. To that aim, the frameworks and regulations can be grouped according to whether their effectiveness has been tested, their scalability, their adaptability, and relevance, whether they take into account industry standards, if they permit multilevel self-declaration as well as third-party certification, the extent of their stakeholder engagement; the coordination of different stakeholders; and their compliance with harmonization among different privacy and security regulations [35, 36]. In addition, informational materials have been explored in order to check the relevance of the collected details and to elicit broader insights into the groups and frameworks.

4.5 DESIGN AND IMPLEMENTATION OF AIoT SECURITY GOVERNANCE FRAMEWORKS

AIoT security governance frameworks combine a set of security practices to help secure interconnected devices. They aim to align the security available in the connected IoT devices. First, consider the best practices for designing a robust structural and process-based AIoT security governance framework. The road to a successful and all-inclusive governance framework follows a path of participation, change, and comprehensive coverage by relevant stakeholders. A security governance framework must meet the requirements of the adopters, who are the intended consumers of this framework. Consequently, relevant stakeholders such as manufacturers, policymakers, young people, health professionals, IT vendors, and others will be engaged to collaboratively draft the requirements and the critical areas that can be used as building blocks for the AIoT security governance framework [37].

Practical aspects such as implementation and sustainability of governance frameworks are regarded crucial. Some of the inherent good practices pertaining to the design of the appropriate implementation strategies and monitoring for exceptions and modifications are discussed for the benefit of novice developers of governance frameworks. For a good practice to be truly effective, a feedback loop for continuous improvement must also be embedded in these governance frameworks [38]. When different national and international organizations or working groups, each advocating their set of governance observance, appear operational, propositions based on different visions and values would likely creep into society, which could probably impede the development of inclusive social relations development. Communication must be established. Governance frameworks need to be written down [39]. They must be systematically managed and subject to surveillance, audit, and review, and where indicated, they must be revised and updated. All policies and procedures in the framework need to be communicated appropriately to every person who has a role related to the agreement if, indeed, a wide, confident, and relationship-sustaining provider is intended [18].

4.5.1 Key components of a governance framework

The key components that constitute the key areas of an AIoT security governance framework include, but are not limited to: Policy development to support security priorities and ensure the security, safety, data integrity, and privacy of staff. Risk assessment includes looking forward to the potential opportunities of using secure AIoT, as well as potential threats. Compliance monitoring: monitoring and reporting compliance with AIoT security and safety standards and ensuring regular review and updates. Operational frameworks to manage assaults, security, and privacy breaches [5, 17]. Stakeholder collaboration: acting together within priority developments to agree on what to work on without affecting our key strategic interests. Best practices include a template and a guideline on:

- Governance framework for AIoT security development, uptake, and evaluation monitoring equipment.
- Definitions: in the first instance, a definition of terminology is to be developed in line with the outlines.
- Minimum security management (both physical and virtual) for devices operating within the Critical National Capability.
- Surveillance: to develop the minimum requirement for security management for property surveillance.

AIoT security governance should align with the organization's objectives and quality plans. The following components, which together form a framework for governance, are all interdependent and continue to build.

4.5.2 Best practices for implementation

The implementation of AIoT security governance frameworks should begin with thorough planning. Engage with stakeholders who may be affected by the governance arrangements and those who are best equipped to articulate business impacts and outlines, as well as other security stakeholders. As part of the planning process, determine the allocation of necessary resources, such as whether dedicated staff will be required. Establish clear roles and responsibilities and a timeline to guide periodic reviews and facilitate any necessary updates [3, 4]. Disseminate the governance arrangements widely across the organization and communicate to the business about decisions that affect their areas. Allocate appropriate resources to facilitate the implementation of AIoT security governance frameworks in your organization. Resources are likely to include sufficient ICT systems and personnel. As with the issue of skills and knowledge, address this matter in your organization in the first instance as part of the implementation process. Work is already underway to identify potential training and capacity-building activities that will foster a security-aware culture. Where an organization has implemented AIoT security governance frameworks, it should establish regular oversight and review. Structures should be regularly evaluated for their effectiveness and to ensure that they stay up to date. Strong emphasis should be placed on documentation [1, 2]. This means that all governance arrangements and actions should be documented and that dialogue should be included in this documentation. This will create a "gold standard" as well as offering a clearer path for other organizations to follow. Engage with possible case studies and discuss the journey they took to improve and create the AIoT security governance framework. They must adopt best practice solutions, as they may have already overcome challenges. You should coordinate the event as an observer, listen to suggestions and challenges, and encourage the organization to report the entire network [1].

4.6 REGULATORY AND ETHICAL CONSIDERATIONS IN AIoT SECURITY GOVERNANCE

Regulatory Considerations AIoT systems must comply with different standards depending on their regions of operation, which, in turn, can influence the perceived AIoT trustworthiness even in regions where no specific regulation applies. There are legal frameworks with which AIoT environments must comply. Failing to meet relevant regulations poses both legal and ethical risks that need to be dealt with. The increasing speed at which data is processed, and flows add complexity to AIoT security governance. Therefore, ensuring conformance to these laws is crucial to avoid liabilities. Different regulatory requirements may also have consequences for how data handling and storage are expected within AIoT systems and hence bear relevance in

AIoT governance [18, 39]. Ethical Considerations: The ethical principles that AI must comply with are used to complement regulatory requirements in the guidelines, accommodate a global perspective and encompass high-risk AI. There are several commonalities between ethical principles and regulatory requirements, from which one can argue that adherence to ethical principles can be regarded as basic governance practice within the regulatory landscape seen through the morals-versus-law tradition. Setting a common global ethic fosters AI innovation and will help in building global trust, thereby enhancing the acceptance and uptake of high-quality AI products [26, 27]. Disparaging views across different countries on what constitutes ethical AI could, in turn, hamper the development of a globally accepted AI innovation ecosystem. The extent to which the regulatory framework should control AI is an ongoing debate. It is important to allow for the unpredictable and exponential growth of AI and not stifle innovation through overly strict regulations. It is also important to recognize the close alignment between many regulations and existing industry-led ethical AI principles in AIoT. Practices consistent with ethical AI principles are already being adopted in AIoT developments [29]. Excellence in governance, driven by adherence to ethical principles, bolsters user trust while ensuring conformance to regulatory requirements. It becomes, therefore, important to balance the demands of regulatory compliance, innovation, and embedding user trust and ethical perceptions in governance approaches. Transparency, audibility, and accountability are guiding lights for AI developers, AI product development, and subsequent interactions with the AIoT. These form a layer that elevates an AIoT framework toward being regarded as ethically aligned. While this may at first invariably lead to some added costs due to technology and process strengthening, audit, and adaptability, it is expected that there could be corporate uptake and consumer preference for AIoT solutions that can make a case for working within such a framework [34, 35].

4.6.1 Legal frameworks and compliance requirements

The development and use of AIoT will be subject to laws and regulations that present established requirements and legal standards that the responsible entities must meet to ensure that these systems are developed and used in a secure, safe, ethical, and trustworthy manner. In order to be compliant with legal frameworks, organizations must redirect their security strategy's main purpose from a pure security coping process to a required process aimed at ensuring that the developed IT security technologies and security-guaranteed use of systems are compliant with the laws, ethics, and regulations [38]. Data Protection and Privacy Laws: In this section, we provide a background to the laws and regulations developed for data protection and privacy, discuss the legal aspects around geographical variations of these laws, and provide an overview of the upcoming data protection laws.

As almost every application of AI/profiling also includes the collection and processing of personal data, AI-responsible entities are also subject to data protection laws and regulations. In the European Union, the General Data Protection Regulation applies to personal data processing and introduces principles, obligations, rights, and liabilities that companies must comply with in order to avoid financial penalties incurred by violations. In addition to the European Union's regulation, different data protection laws and directives apply based on the AI-responsible entity's geographical location. In the United States, personal data protection regulations differ from state to state, while there is no overarching federal data protection law [17, 37]. This results in state-specific territorial compliance requirements, and multinational organizations with headquarters in California that are compliant with local regulations may not be compliant with data protection regulations in other states. As such, it is essential to consider regional variations in data protection laws and directives to ensure that personal data processing is legally sound. Failure to comply with these data protection laws can lead to heavy punishments in terms of fines and legal costs and may impact user trust and confidence in AI [2].

4.6.2 Ethical principles in AIoT

To ensure security, AIoT systems and processes must reflect a series of ethical considerations. Among these is the notion of "data privacy," referring to an organization's ethical obligations regarding what information is collected about customers and how it is used. This, in turn, is tied to the idea of "informed user consent" and the importance of transparency and communication between AIoT operators and the consumers who will be affected by their technology [1]. Just as AIoT organizations ought to respect resource owners through securing the products of their ingenuity, a process premised on trust, so too do these firms have a moral duty to remain "accountable" for their technology in order to maintain public trust. Ethically, AIoT organizations can also be expected to consider the implications of automated decision-making on such ethical principles as "fairness," "transparency," and "non-discrimination" (Shen et al., 2023).

The first consideration when discussing ethical principles in AIoT is fairness. One clear ethical application of the principle of fairness when it comes to AIoT technology is the concept of fighting bias. This will be of particular importance when discussing AIoT devices and data processes used to make human resources decisions, such as resumes or CV scanning tools. Countries with robust discrimination laws might consider the principle of non-discrimination to be of equal importance to fairness in ethical considerations. Similar to other decision-making systems that are able to process data, AIoT systems must also take the principle of transparency into account, especially as it relates to data processing and use [35, 36]. This ethical point considers the use of AIoT systems' "degree of explanation," or the level of user understanding an individual has about the system and how it works. Ethical principles in AIoT strategic goals themselves must also properly navigate regulations in order to comply with them. Therefore, a summary of both ethical and legal compliance is necessary to provide an appropriate response to a user's question. While regulatory and ethical considerations can be nuanced, drafting these guidelines into a recognizable framework that provides a course of action for decision-makers is a necessary and achievable task. Experience from case reports could be useful to support this [31, 38]. Also, there is a growing body of literature supporting the ethical implications of data processing and data control, which may be similarly useful in shaping a coherent governance framework.

Although ethical considerations are not the principal concern for many public regulators, it is nonetheless becoming increasingly recognized not only that the ethical implications of AIoT cannot be ignored but also that promulgating ethical guidelines could be a way of building public trust in AI applications, given rapidly growing public interest in and concern about the technology. As such, ethical guidelines formulated by AIoT organizations must try to strike a balance between legal compliance and ethical considerations [27]. To this end, it makes sense to bring together ethical considerations with a basis in existing legislation according to the jurisdictions in which the organization currently operates to form a "best practice" framework. It should be noted as a caveat that legal terms are linked to the framework of the GDPR. Recognizability is not the standard for legal compliance, as policies vary according to the state [6, 7].

4.7 CASE STUDIES AND PRACTICAL APPLICATIONS

To explore the practicality of AIoT security governance, five case studies are presented to illustrate where AIoT deployments pose significant challenges, as well as the corresponding methods of governance. They demonstrate the range of governance challenges within specific AIoT deployments and how they have been addressed, leading to improvements in the security posture and risk management of the organization that deployed them. These advances in risk management have led to increased confidence, trust, and outcomes that have potentially mitigated operational risk that was threatened by an irresponsible or high-impact event. Governance strategies employed in each case study are discussed to demonstrate both the specific learnings from these real-world cases and to showcase the importance of the context in AIoT security governance [14, 15]. Each case study introduces the AIoT deployment, organization, and governance plan that was implemented in response to a significant governance challenge. The case studies selected for inclusion consider AIoT deployments in large and small multinational companies, large banks, infrastructure operators, and a connected living concept. These deployments cover the spectrum of normal and novel

AIoT use, which routinely occur in society today. These are not case studies of a future we have not yet entered, but rather, they use current and future solutions that are either available or able to be delivered today to enable an innovative real-world deployment [16, 23]. These case studies have been structured to demonstrate AIoT security governance from a practitioner's perspective, which is reflected in their practical challenges and possible governance responses. The practicalities and outcomes of these AIoT deployments are explored in detail to present the real-world challenges of cyber connectivity and the consequent need for effective AIoT security governance solutions. Such AIoT security governance solutions are unique to the specific challenge, considering the organization's operations, people, and clients, and are influenced by the wider political, economic, societal, technological, legal, and environmental context. These case studies are focused on how useful the strategic knowledge of the developments in AIoT technologies and agencies are in practical deployments of AIoT in the private sector [22].

4.7.1 Real-world examples of AIoT security governance

A comprehensive IoT security program consists of five key building blocks that together address the entirety of an IoT ecosystem in use cases. The IoT program is built upon a more general information security program, where IoT collects registration and notification of all IoT assets recognized as part of the framework. Initially, there were 130 devices registered for 590 users/owners. This number changes daily as IoT adoption continues apace. The security posture and risk rating of these assets are tracked in near real-time via a reporting algorithm [20]. The framework is thus in constant improvement as it is reassessed in response to emerging and evolving security threats, maintaining a strong investment in governance processes of cyber risk. A large-scale collaborative initiative is designed to create a high-assurance, secure, and open-source stack for the AIoT. The project focuses on leveraging recent innovations for security and related controls to deliver robustness requirements appropriate for maintained AIoT ecosystems [25]. Collaborative stakeholder-driven research engagements provide overviews of how to build secure systems; the current approach bypasses existing insecure architectures and presents practical examples of good systems. Fieldwork was conducted as part of the initiative. A field study highlighted a small set of successful emerging governance practices in this context. A small selection of these practices is discussed in the following subsections. However, it is also worth noting that a number of mechanisms were found to be effective in leveraging investment in AI to improve the security of these devices by making initial default configurations and responses to anomalous behavior more secure [22, 23]. An example of an overarching governance framework was seen in a hospital IT department used to govern security in applications.

4.7.2 Lessons learned and success stories

This brings us to the point of opportunities and challenges to governing AIoT security. In the following, we summarize the lessons learned, emphasizing the ability to relate them to practical actions for other organizations to address and offering advice on doing so. Based on the lessons learned and observations, we derived two success stories from HealthLake Cloud and NAKIVO. We examined eight cases of AIoT security governance frameworks in organizations that received preparatory action for standards from the current expertise in the marketplace [19]. Organizations came from the automotive, healthcare, energy, and environmental sectors; the majority of organizations were suppliers of advanced equipment, tools, and parts. Some cases are still in iterative developmental steps to streamline all the resources for deployment. The cases took into account a range of risks, such as adverse impacts on safety, data abuse, financial loss, data loss, unauthorized intrusion into devices, service loss or corruption, and outright sabotage. The results from the cases led us to identify lessons learned that can be used to improve future efforts to protect AIoT solutions [26].

As it was with Welfare and CHI, HealthLake Cloud is fairly new. However, because it is already embedded in Amazon Web Services, other organizations are building infrastructure using HealthLake. Therefore, some of the security KPIs from AWS are all needed for HealthLake. These include access control, data encryption at rest and in transit, and encryption key management. In contrast to the above, NAKIVO has released comparative cybersecurity illumination for its multicloud support. As NAKIVO ensures that cloud and IoT solutions on-premises can store data as they wish, all information uploaded to NAKIVO infrastructure is treated as a secure connection [27, 28]. They conduct user authorization and encryption to achieve the key security objective. In the case studies, AIoT solutions like HealthLake and NAKIVO embed technologies based on AI from some innovative suppliers. These technologies have been iteratively assisted and enhanced by leading researchers and manufacturers. Despite the recent grounding process, the suppliers embed many safety functions to attract prospective customers [30, 40]. All the suppliers continue to diversify how and to whom they are marketing as they complete the trial of AIoT.

4.8 FUTURE TRENDS AND EMERGING TECHNOLOGIES IN AIoT SECURITY GOVERNANCE

The more integrated AIoT extends across sectors and services, the more reliable and secure these technologies need to be. This means that trends in the development of AIoT technologies will likely have fundamental implications for the future of security governance frameworks. Ongoing developments in machine learning and big data analytics are expected to further strengthen

AIoT security [7]. As algorithms improve and the volume of available data increases, the accuracy of these AI-based security systems in threat detection is anticipated to grow. Consequently, achieving consensual data and algorithmic governance will likely be high on future agendas. Both national and international regulatory standards will have to adapt to deal with new technologies, rather than ban or devalue them. New technologies such as blockchain for transactional data or edge computing also aim to improve the security of AIoT devices [10]. Governance frameworks developed today need to be flexible, opening up to future, continuous adjustments and innovations in light of constantly advancing security technology, even if their targets, principles, or definitions are rooted in the current state of the art [23]. As threatening and uncertain some future AIoT trends may appear to be, embedding smart logic and security into AIoT hardware is still only taking its first steps. As the AIoT evolves, organizations must keep an eye on developing trends as they prime opportunities for governance innovation and address corresponding ethical and security needs with proactive governance strategies [19, 20].

4.8.1 Predictions for the future of AIoT security governance

AIoT systems are expected to evolve and continue to form the core of complex critical infrastructures in smart cities, smart homes, and smart factories in the future. Based on this and the findings of recent studies, the following aspects may become particularly relevant in the years to come: The increasing complexity of systems will drive AIoT security research toward more agile, adaptable, and scalable governance frameworks [25]. AI can be employed for predictive and proactive security monitoring and automated real-time decision-making to take remediation actions based on the findings of IoT security analytics. This could help in automating the governance functions and real-time adaptation of AIoT security processes to changes (for instance, in requirements, environments, malfunctions, and threats). Moreover, AI could be used to optimize the distribution of governance tasks based on the required effort, security guarantees, and trustworthiness of security-relevant devices. The rapid development of AI algorithms will lead to an increase in the level of complexity such that it can be predicted, but only automated and embedded via software development kits and toolkits in smart space deployments and IoT devices [27, 28].

In the near future, data-driven regulatory and certification priorities will shift to AIoT security as an outcome of AI, IoT, and multidisciplinary governance, without a radical rethinking of the appropriateness of IoT and AI regulations and governance. This perspective leaves a spot for discussions about the limitations of ad-hoc IoT technology regulations as an optimal track to address AIoT governance. In order to counteract current trends, a collaborative approach of stakeholder-oriented governance as a process is essential. In parallel, the AI deep tech revolution will create new patterns in security-related governance [40]. Emerging regulations could have a major influence on the development and deployment of industry-driven AIoT security experts who can provide the professional skills needed to ensure that they are built as a standard from the beginning of development. Moreover, the underlying security-oriented conventional certifications are an opportunity that primarily addresses liabilities and widespread effects issues in order to provide better "beyond strictly required" practices. There are now initiatives to provide powerful software for the increasingly widespread problem of standards and certifications for AIoT systems [36]. In addition, the organization of the standards' subsequent adaptations in this constantly evolving game changer is needed. For the next few years, organizations, associations, government institutions, and other international actors, including stakeholders, are needed to integrate these new governance requirements. This work can be achieved in part through adaptation, continuous learning, and experience that combines all operators' real expertise in AIoT, security, and regulation, through a multistakeholder surveillance process and center [18].

4.8.2 Impact of emerging technologies

Emerging technologies such as blockchain, quantum computing, edge computing, advanced analytics, and other interoperable communication infrastructure innovations can help enhance AIoT security. As advanced computing has given rise to various AIoT systems, the use of quantum computing can create the power to crack encryption algorithms. Advanced computing systems enabled by quantum computing can help secure AIoT systems. However, these technologies also introduce new attack vectors, techniques, and vulnerabilities; therefore, closely integrated and intertwined systems create new dependencies, additional levels of complexity, and further governance implications [35]. Governance and policy frameworks should consider building security resilience in accordance with the advantages and challenges posed by new and emerging system technologies.

Resilient governance frameworks are systems of governance and sociotechnological ecosystems that have the potential to recover, evolve, and adapt to the new technological evolution. For instance, blockchain offers decentralized security storage, ensuring that the data in an AIoT system is secured. Blockchain can bring about a transformation in reducing security threats to data, which in turn can protect the privacy of customers. However, the integration of emerging technologies in AIoT systems increases the interdependencies between the governance structure of existing systems and the embedded infrastructure [18]. Every developing and emerging technology must be grafted into existing systems with prior knowledge of innovations. The continuous evolution of technology has enabled the development of loopholes in the system and sparked technological innovation discussions in the governance framework. Moreover, newer technological discussions provide gaps that cannot be administered and managed with existing policies. Emerging or developing technologies offer investment opportunities for robotics, autonomous vehicles, IoT systems, and resilient AIoT-connected devices [1, 32].

4.9 CONCLUSION

In recent years, advancements in machine learning have resulted in zero-day attacks, where advanced attacks are executed dynamically without the creation of known patterns. These attacks can be launched on low-cost devices due to the advent of artificial intelligence-optimized programming languages. Also, with the increasing capability of IoT devices, innovations in blockchain and AI tandem, the AIoT framework, have gained momentum. These emerging technological trends point to reflective implications in AIoT security governance, ensuring AIoT security is a challenging task due to the dynamic and unpredictable nature of attackers. Future work must incorporate an adaptive AIoT security governance model that takes into account rapid technological advancements in AI, IoT, and big data. As we highlight, AIoT needs a collaborative effort among industry, governmental entities, policy and law, non-governmental organizations, and academia, in addition to the classic asset and technology players in a socio-technical environment. Hence, there is a need to develop a governance model that is not only adaptive to tackle AIoT-specific attacks but also takes into account the multitude of stakeholders in play across these ever-challenging, connected technologies. Importantly, one thread throughout this research has highlighted the growing need for ethical governance frameworks in AIoT, addressing the challenges associated with the ingesting of big data that inevitably comes with a breach of users' privacy and data portability. With the excitement in the capabilities and capacities ascribed to a sovereign IoT, little attention, to date, has been given to the potentially damaging consequences of not adhering to best practices and sound governance. As we can see now, security governance in AIoT did not stand still, and this is a valuable observation—and a call for future work.

REFERENCES

[1] Pise A. A. et al., "Enabling artificial intelligence of things (AIoT) healthcare architectures and listing security issues," *Computational Intelligence and Neuroscience*, vol. 2022, 2022, doi: 10.1155/2022/8421434

[2] Liu W. et al., "D2MIF: A malicious model detection mechanism for federated-learning-empowered artificial intelligence of things," *IEEE Internet of Things Journal*, vol. 10, no. 3, 2023, doi: 10.1109/JIOT.2021.3081606

[3] Wazid M., Das A. K., and Park Y., "Blockchain-envisioned secure authentication approach in AIoT: applications, challenges, and future research," *Wireless Communications and Mobile Computing*, vol. 2021, no. 1, 2021, p. 3866006. doi: 10.1155/2021/3866006

[4] Aliahmadi A. and Nozari H., "Evaluation of security metrics in AIoT and blockchain-based supply chain by Neutrosophic decision-making method," *Supply Chain Forum*, vol. 24, no. 1, 2023, doi: 10.1080/16258312.2022.2101898

[5] Yi P. and Li Z., "Construction and management of intelligent campus based on student privacy protection under the background of artificial intelligence and internet of things," *Mobile Information Systems*, vol. 2022, 2022, doi: 10.1155/2022/2154577

[6] Chang K. J., Chuang C. W., Chiu J. T., and Chen J. Y., "Flying watchdog: A drone with edge AIoT for residential safety and fall detection by face and posture recognition," *in APWCS 2022 - 2022 IEEE VTS Asia Pacific Wireless Communications Symposium*, 2022. doi: 10.1109/APWCS55727.2022.9906504

[7] Shafik W., "IoMT future trends and challenges: Emerging technologies, policy implications, and research questions," *Lightweight Digital Trust Architectures in the Internet of Medical Things (IoMT)*, IGI Global, pp. 348–470, 2024, doi: 10.4018/979-8-3693-2109-6.ch019

[8] Wang P., "Digital multimedia signal processing via AIoT and its application in smart home," *Internet Technology Letters*, vol. 6, no. 5, 2023, doi: 10.1002/itl2.335

[9] Han W., Peng J., Yu J., Kang J., Lu J., and Niyato D., "Heterogeneous data-aware federated learning for intrusion detection systems via meta-sampling in artificial intelligence of things," *IEEE Internet of Things Journal*, vol. 11, no. 8, 2024, doi: 10.1109/JIOT.2023.3337755

[10] Muhammad Adnan Khan, "An intelligent and secure communication of AIoT enabled devices empowered with IPK algorithm," *Lahore Garrison University Research Journal of Computer Science and Information Technology*, vol. 3, no. 4, 2019, doi: 10.54692/lgurjcsit.2019.030487

[11] Shafik W., "Artificial intelligence-enabled internet of medical things (AIoMT) in modern healthcare practices." In *Clinical Practice and Unmet Challenges in AI-Enhanced Healthcare Systems*, IGI Global, pp. 42–69, 2024, doi: 10.4018/979-8-3693-2703-6.ch003

[12] Alaba F. A., Jegede A., Sani U., and Dada E. G., "Artificial intelligence of things (AIoT) solutions for sustainable agriculture and food security," *in Lecture Notes on Data Engineering and Communications Technologies*, vol. 192, 2024. doi: 10.1007/978-3-031-53433-1_7

[13] Shafik W., "The future of healthcare: AIoMT—redefining healthcare with advanced artificial intelligence and machine learning techniques," *Artificial Intelligence and Machine Learning in Drug Design and Development*, vol. 12, pp. 605–634, 2024, doi: 10.1002/9781394234196.ch19

[14] Wu Y. C., Wu Y. J., and Wu S. M., "An outlook of a future smart city in taiwan from post-internet of things to artificial intelligence internet of things," *in Smart Cities: Issues and Challenges Mapping Political, Social and Economic Risks and Threats*, 2019. doi: 10.1016/B978-0-12-816639-0.00015-6

[15] Zhuang Y., Wang C., Zheng W., Victor N., and Gadekallu T. R., "ERACMA: Expressive and revocable access control with multi-authority for AIoT-enabled

human centric consumer electronics," *IEEE Transactions on Consumer Electronics*, vol. 70, no. 1, 2024, doi: 10.1109/TCE.2023.3306752
[16] Wang Y., Liang X., Hei X., Ji W., and Zhu L., "Deep learning data privacy protection based on homomorphic encryption in AIoT," *Mobile Information Systems*, vol. 2021, 2021, doi: 10.1155/2021/5510857
[17] Shafik W., "Artificial intelligence-enabled internet of medical things for enhanced healthcare systems," *Smart Healthcare Systems*, pp. 119–134, 2024, doi: 10.1201/9781032698519-9
[18] Tan L., Yu K., Ming F., Cheng X., and Srivastava G., "Secure and resilient artificial intelligence of things: A HoneyNet approach for threat detection and situational awareness," *IEEE Consumer Electronics Magazine*, vol. 11, no. 3, 2022, doi: 10.1109/MCE.2021.3081874
[19] Bao H. et al., "A probabilistic and distributed validation framework based on blockchain for artificial intelligence of things," *IEEE Internet of Things Journal*, vol. 11, no. 1, 2024, doi: 10.1109/JIOT.2023.3279849
[20] Shafik W., "Connected healthcare—the impact of Internet of Things on medical services: Merits, limitations, future insights, case studies, and open research questions," In *Artificial Intelligence and Internet of Things based Augmented Trends for Data Driven Systems*, pp. 181–217, 2024. CRC Press, doi: 10.1201/9781003497318-10
[21] Leong Y. M., Lim E. H., Subri N. F. B., and Jalil N. B. A., "Transforming agriculture: Navigating the challenges and embracing the opportunities of artificial intelligence of things," in *2023 IEEE International Conference on Agrosystem Engineering, Technology and Applications, AGRETA 2023*, 2023. doi: 10.1109/AGRETA57740.2023.10262747
[22] Shafik W., "Smart health revolution: Exploring artificial intelligence of internet of medical things," In: Kumar, P., Singh, P., Diwakar, M., Garg, D. *(eds) Healthcare Industry Assessment: Analyzing Risks, Security, and Reliability. Engineering Cyber-Physical Systems and Critical Infrastructures*, vol 11, pp 201–229, 2024. Springer, Cham, doi: 10.1007/978-3-031-65434-3_9
[23] Jia Y., Lin F., and Sun Y., "A novel federated learning aggregation algorithm for AIoT intrusion detection," *IET Communications*, vol. 18, no. 7, 2024, doi: 10.1049/cmu2.12744
[24] Ye L. et al., "The challenges and emerging technologies for low-power artificial intelligence IoT systems," *IEEE Transactions on Circuits and Systems I: Regular Papers*, vol. 68, no. 12, 2021, doi: 10.1109/TCSI.2021.3095622
[25] Shafik W., "Healthcare and its applications in a smart city," *Healthcare-Driven Intelligent Computing Paradigms to Secure Futuristic Smart Cities*, vol. 1, 2024, doi: 10.1201/9781032631738-1
[26] Mei Y., Wang W., Liang Y., Liu Q., Chen S., and Wang T., "Privacy-enhanced cooperative storage scheme for contact-free sensory data in AIoT with efficient synchronization," *ACM Transactions on Sensor Networks*, vol. 20, no. 4, 2024, doi: 10.1145/3617998
[27] Chehri A., Jeon G., Rivest F., and Mouftah H. T., "Evolution and trends in artificial intelligence of things security: When good enough is not good enough!," *IEEE Internet of Things Magazine*, vol. 5, no. 3, 2022, doi: 10.1109/iotm.001.2100130

[28] Zhang Z., Zeng K., and Yi Y., "Blockchain-empowered secure aerial edge computing for AIoT devices," *IEEE Internet Things Journal*, vol. 11, no. 1, 2024, doi: 10.1109/JIOT.2023.3294222
[29] Zhang J. and Tao D., "Empowering things with intelligence: A survey of the progress, challenges, and opportunities in artificial intelligence of things," *IEEE Internet of Things Journal*, vol. 8, no 10, 2020, p. 7789–7817. doi: 10.1109/JIOT.2020.3039359
[30] Mika K. et al., "VEDLIoT: Next generation accelerated AIoT systems and applications," in *Proceedings of the 20th ACM International Conference on Computing Frontiers 2023, CF 2023*, 2023. doi: 10.1145/3587135.3592175
[31] Rathee G., Garg S., Kaddoum G., Choi B. J., Hassan M. M., and Alqahtani S. A., "TrustSys: Trusted decision making scheme for collaborative artificial intelligence of things," *IEEE Transactions on Industrial Informatics*, vol. 19, no. 1, 2023, doi: 10.1109/TII.2022.3173006
[32] Qiu X., Yu J., Zhuang W., Li G., and Sun X., "Channel prediction-based security authentication for artificial intelligence of things," *Sensors*, vol. 23, no. 15, 2023, doi: 10.3390/s23156711
[33] Zhang P., Xu X., Qin X., Liu Y., Ma N., and Han S., "Evolution toward artificial intelligence of things under 6G ubiquitous-X," *Journal of Harbin Institute of Technology (New Series)*, vol. 27, no. 3, 2020, doi: 10.11916/j.issn.1005-9113.20036
[34] Padmaja M., Shitharth S., Prasuna K., Chaturvedi A., Kshirsagar P. R., and Vani A., "Grow of artificial intelligence to challenge security in IoT application," *Wireless Personal Communications*, vol. 127, no. 3, 2022, doi: 10.1007/s11277-021-08725-4
[35] Shafik W., "Digital healthcare systems in a federated learning perspective," In *Federated Learning for Digital Healthcare Systems*, pp. 1–35, 2024. Academic Press, doi: 10.1016/B978-0-443-13897-3.00001-1
[36] Wu B. and He S., "Self-learning and explainable deep learning network toward the security of artificial intelligence of things," *Journal of Supercomputing*, vol. 79, no. 4, 2023, doi: 10.1007/s11227-022-04818-4
[37] Sleem A. and Elhenawy I., "Survey of artificial intelligence of things for smart buildings: A closer outlook," *Journal of Intelligent Systems and Internet of Things*, vol. 8, no. 2, 2023, doi: 10.54216/JISIoT.080206
[38] Guo B., Liu S. C., Liu Y., Li Z. G., Yu Z. W., and Zhou X. S., "AIoT: The concept, architecture and key techniques," *Jisuanji Xuebao/Chinese Journal of Computers*, vol. 46, no. 11, 2023, doi: 10.11897/SP.J.1016.2023.02259
[39] Raja G., Essaky S., Ganapathisubramaniyan A., and Baskar Y., "Nexus of deep reinforcement learning and leader-follower approach for AIoT enabled aerial networks," *IEEE Transactions on Industrial Informatics*, vol. 19, no. 8, 2023, doi: 10.1109/TII.2022.3226529
[40] Liu Y., Huang P., Yang F., Huang K., and Shu L., "QuAsyncFL: asynchronous federated learning with quantization for cloud-edge-terminal collaboration enabled AIoT," *IEEE Internet of Things Journal*, vol. 11, no. 1, 2024, doi: 10.1109/JIOT.2023.3290818

Chapter 5

AI-enhanced threat detection and response framework for advanced cyber-physical smart ecosystems

Deepika Malve, Kishor Kumar Reddy C, Meenal H, Pagadala Indira, and Kari Lippert

5.1 INTRODUCTION

"Intelligent environments" basically means connected systems where devices, sensors, and infrastructure are interlinked to make processes more efficient, safer, and even enhance the user experience. Such environments come in the form of smart homes, smart cities, industrial IoT, and healthcare systems that hold aloft the offers of cloud computing, edge computing, and IoT. Cloud computing provides both hardware and software aspects as well as the systems in data centers that contribute to the service [1]. AI and machine learning emerged as powerful tools for assisting diagnosis, determination of the type of prosthesis requirement, the development and positioning of clasps in RPD, designing of connectors and pontics, etc. [2]. With these dangers of developing an increased dependence on digital technology being cited, there should be strong security put in place to protect operational integrity. It underscores the potential applications that can redefine what is achievable, while also addressing the pivotal role of change management in facilitating a smooth transition into this quantum AI-augmented future [3].

Smart environments face several types of threats caused by their connectivity of sensors, IoT, and advances in technologies. In the IIoT philosophy, intelligent machines are not only better at capturing and analyzing data in real time than humans, they can also convey important information that can actually affect the speed and accuracy of decision-making [4]. Cyber threats such as ransomware attacks, DDoS attacks, and even data breaches can compromise sensitive information or disrupt processes operating within the system. Data privacy and security concerns also emerge as formidable barriers to the widespread adoption of AI in the smart economy [5]. Other physical hazards, including tampered sensors or apparatus, might result in system failures with extremely detrimental outcomes. Unauthorized accesses to sensitive user information are a violation of privacy and hence pose considerable risks to individual security as well as organizational security. In applications where it is not possible to provide a stable fixed connection to the Internet, or its installation is complicated and economically inefficient, the way to go is to connect using ultra-fast 5G modules to a high-speed 5th

DOI: 10.1201/9781003606307-5

generation network [6]. Complexity of these ecosystems is posed by large networks and stakeholders, hence demanding new approaches that deploy AI and real-time analytics in ways that deliver safety and resilience.

Artificial intelligence (AI) is revolutionizing cybersecurity by changing how threats are recognized and addressed. Artificial intelligence (AI) enables automation and significantly reduces the need for human engagement by providing real-time monitoring and automatic responses to emerging threats. Artificial intelligence (AI) uses machine learning algorithms to examine trends and anomalies in order to predict potential vulnerabilities before they are exploited. This proactive approach enables organizations to manage threats ahead of time. Additionally, AI systems are highly adaptable, continuously learning from new data and evolving threats, which enhances their ability to bolster systems' security and resilience over time. Machine learning is also known as predictive analytics that makes predictions about certain unknowns in the future through the use of data and is used to solve many real-world business issues, for example, business risk prediction [7]. AI's dynamic adaptability makes it a crucial component of modern cybersecurity techniques. The term "AI on Edge" refers to the execution of AI processes right on edge devices, whereas "AI for Edge" refers to the deployment of AI models and algorithms in the central servers or upper layers to enhance edge computing capabilities [8]. This chapter looks at how AI can improve threat management strategies in next-generation smart environments. By examining important technologies, frameworks, and real-world applications, it shows how AI may change security paradigms and address the unique challenges posed by smart ecosystems.

5.2 CORE CONCEPTS AND TECHNOLOGIES

Fundamentals and tools: AI-enhanced threat management in intelligent environments relies on the technologies and approaches that allow systems perceive, counter, and resolve cyber and physical threats. Machine learning algorithms, fundamentally at the core of these systems, process large amounts of data being generated through devices interlinked and then extract patterns, anomalies, or possible vulnerabilities. The same methodology often employs supervised and unsupervised learning approaches. Supervised models are trained on labelled datasets and are designed to find the known threats, while unsupervised models work on the aberrant behavior for discovering new or emerging threats. The efficiency of the algorithm greatly depends on a good dataset that is varied and not heavily skewed [9]. Network resilience, low-latency networking solutions, network slicing, and virtualization are key enablers for robust connectivity, and 5G and future telecommunication standards will unlock new potential for remote connectivity and service delivery, as outlined in the research study by McMahan et al. [10]. Natural language processing, or NLP, is also used to analyze the information in textual form derived from security logs and feeds of threat intelligence for the assistance of systems to understand and respond to events related to security. The processed satellite images contained features like rain, snow, Tropical depression (T.Depression), thunderstorms (T.strom), and cyclone [11]. From the cloud, they identify a variety of cybersecurity issues, including system and application vulnerabilities, malware injection attacks, denial of service (DoS), malicious insider threats, and data leakage [12].

Table 5.1 outlines key AI technologies such as machine learning, anomaly detection, and edge computing, along with their specific applications in threat management for smart environments.

A significant improvement toward the field is represented by edge computing, which permits the location of AI functionalities close to sources of environmental data – such as sensors in IoT – in order to extract real-time threats against latency; by this way, it can avoid the need to transmit all the data to centralized servers. Advances in virtualization have paved the way for the emergence of Internet clouds as a novel paradigm [13]. Within intelligent environments, AI is combined with security protocols in IoT, including TLS, MQTT, and CoAP, to make sure that secure communication between devices and central systems exists. Anomaly detection is another

Table 5.1 Core concepts and technologies in AI-augmented threat management

Core concept/ technology	*Description*	*Application in threat management*
Machine Learning (ML)	Algorithms that learn from data to identify patterns and make predictions.	Used for anomaly detection, identifying known and unknown threats.
Anomaly Detection	Identifying deviations from normal patterns.	Detects unusual behavior, potential cyberattacks or physical threats.
Edge Computing	Processing data closer to the data source, reducing latency.	Enhances real-time threat detection and response in distributed systems.
Natural Language Processing (NLP)	Analyses and interprets textual data such as security logs.	Enables understanding of threat intelligence feeds, logs, and alerts.
Predictive Analytics	Analyses historical data to forecast potential threats.	Anticipates vulnerabilities, reducing response times and enabling proactive defense.
IoT Security Protocols	Protocols ensuring secure communication between IoT devices.	Safeguards data integrity and secure connections between devices in smart environments.

important technique that can be used to identify any anomalies deviating from normal system behavior, which might indicate a cyberattack or some flaw in the physical infrastructure. The evolving smart city applications running on the underlying 6G networks require high reliability and high security. In this context, intrusion detection can be used to identify unauthorized access and malicious activities in smart city applications [14]. Another tool used by AI systems is predictive analytics, which uses previous data to foresee potential threats and, hence, supports proactive threat management techniques. Altogether, these technologies form a robust framework for protecting intelligent environments. AI-augmented security frameworks must be based on such fundamental principles that marry aspects of building integrated systems capable of independently identifying risks, assessing them, and mitigating them. Insofar as progress augments systems to also learn and evolve, they must become capable of identifying new threats to achieve truly scalable, adaptable defenses necessary for the ecosystems that define next-generation intelligent environments (Figure 5.1). The fact that quantities of data that would not have been collected had the AI not been employed as part of intelligence analysis is where the dangers for increased levels of intrusion lie [15].

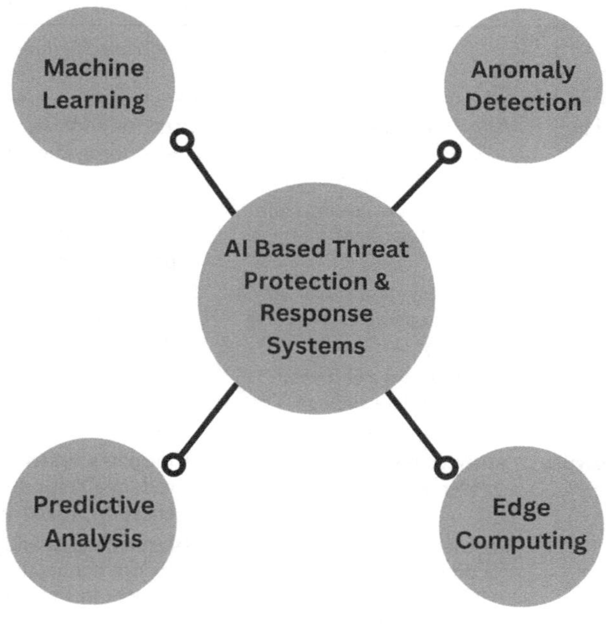

Figure 5.1 Key components of AI-based threat protection and response systems.

5.3 AI-POWERED THREAT DETECTION

To promptly identify and mitigate security vulnerabilities within intelligent environments, an array of advanced machine learning methodologies is integrated into AI-driven threat detection systems. This functionality relies on techniques for anomaly detection, which consistently oversee network traffic, device interactions, and system behavior to detect divergences from established norms. These anomalies may be indicative of more savvy cyber-attacks like malware infections, ransomware, or exfiltration of data and even physical threats from corrupted IoT sensors or devices. In the complex landscape of AI-infused IoT systems, transparency and interpretability are pivotal qualities for informed decision-making and effective governance [16]. Unlike signature-based traditional methods requiring known danger patterns, AI systems learn using unsupervised learning to look for unknown threats through identification of unusual patterns or behavior that are no longer consistent with past norms. The double approach brings anomaly detection along with signature-based defense mechanisms into a single system for a robust protection against recognized and unknown attack vectors. Moreover, systems with artificial intelligence continue to improve and enhance their detection ability through machine learning and predictive analytics.

When AI models are trained to identify the changing TTPs used by hackers through learning from historical data and taking feedback from previous incidents, threat detection speed, and accuracy increase. For example, predictive models will be able to predict probable vulnerabilities or attack routes by identifying patterns that precede breaches in security. This gives businesses a chance to address risks before they get exploited. The detection of small, low-level threats remains to be a significant area in which these artificial intelligence systems become increasingly effective as, through their ongoing learning behavior, they strengthen their defense mechanism against both new and sophisticated attacks. Moreover, using NLP improves AI's ability to process logs, security feeds, or any other unstructured data, enabling them to understand the threats in greater depth and with more sophistication.

Table 5.2 provides an overview of various AI-based threat detection methods, including signature-based detection, anomaly detection, machine learning, and predictive analytics.

A third dimension of artificial intelligence's capability in threat identification is the ability to respond in real time. After detecting the threat, AI systems can immediately take automated action to counteract malicious communications by closing compromised devices, cutting off malicious communications, or alerting human responders. Real-time response to attacks is critical because widely distributed DDoS attacks and big data breaches cannot be mitigated once they are executed. IoT devices rely on advanced communication protocols and networks to share the acquired data in real time. Leveraging the potential of advanced sensing and communication abilities, significant efforts are being made to revolutionize the IoT

Table 5.2 AI-powered threat detection methods

Method	*Description*	*Use case*
Signature-Based Detection	Identifies known threats based on predefined patterns.	Detects well-known malware and attack signatures.
Anomaly Detection	Identifies deviations from baseline behaviors, enabling detection of unknown threats.	Detects novel threats by recognizing abnormal patterns in device behavior or network traffic.
Machine Learning (Supervised and Unsupervised)	Supervised learning uses labelled data, while unsupervised learning detects patterns without labels.	Supervised: Identifies known attacks. Unsupervised: Detects emerging or unknown threats.
Predictive Analytics	Uses historical data to forecast potential threats before they manifest.	Prevents attacks by identifying vulnerabilities before exploitation.

experience further [17]. AI-enhanced security frameworks that also have an easy interface with edge computing as well as IoT security standards allow decentralizing the process and enables faster decision-making at the point of attack. Real-time analysis, automated remediation, and continuous learning by an AI-powered threat detection system provide a proactive robust defense against the myriad physical and cyber threats in the Smart Environment by adopting a flexible approach toward scalable security (Figure 5.2).

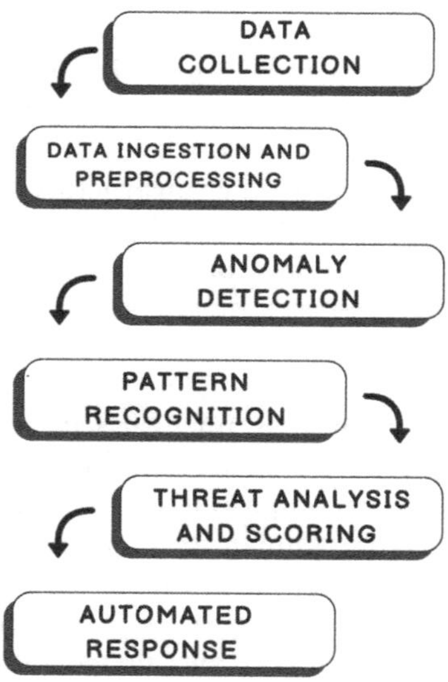

Figure 5.2 AI-based threat detection workflow.

5.4 FRAMEWORK FOR AI-AUGMENTED THREAT MANAGEMENT

The architecture developed for smart threat management contains quite a few advanced technologies and methodologies used to protect and strengthen the resilience of cyber-physical systems. Specifically, it makes use of machine learning techniques and applies both supervised and unsupervised learning in processing the enormous amount of data generated by sensors, network traffic, and other Internet of Things (IoT) devices. Augmented reality technology allows for explanations of procedures going beyond mere theoretical knowledge [18]. These AI models, in collaboration with anomaly detection software, find and isolate probable dangers, such as ransomware, physical intrusions such as sensor hacking, and privacy invasions like data access in violation of privacy, through the study of anomalies in normal patterns. Federated learning (FL), which bases decisions on the local dataset, can avoid data breaches and help with privacy maintenance because decisions can be made without recourse to any central server [1]. Artificial intelligence provides a better adaptive defense as compared with the prevailing security devices since it helps process data in real time and detect known and unknown threats simultaneously.

Table 5.3 presents the key components of an AI-driven threat management framework, detailing their functions and the technologies involved.

Flexibility is one of the most notable features of the AI-infused threat management system. Artificial intelligence systems are designed to assimilate insights from new data and past incidents, making them better over time to identify and respond. Predictive analytics aids in adopting proactive threat management by allowing artificial intelligence to scan the past data for developing patterns or potential vulnerabilities. Other roles include edge computing, which makes it possible to analyze data near the source to

Table 5.3 AI-augmented threat management framework components

Component	*Function*	*Technology involved*
Threat Detection	Identifies potential cyber and physical threats.	Machine Learning, Anomaly Detection
Real-Time Response	Automates threat mitigation actions in real-time.	Automated Security Systems, Edge Computing
Predictive Threat Management	Forecasts potential threats to enable proactive defense.	Predictive Analytics, Machine Learning
Data Privacy and Integrity	Ensures secure communication and privacy of sensitive data.	IoT Security Protocols, Blockchain
Continuous Learning	Adapts and improves over time based on new data.	Federated Learning, Machine Learning

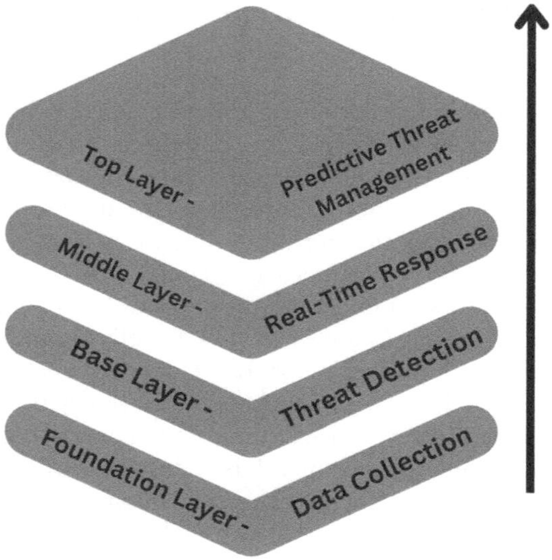

Figure 5.3 Layered architecture of AI-based threat management systems.

reduce latency and support real-time capabilities in decision-making. This decentralized way ensures AI-enhanced security frameworks correctly work even under loosely connected conditions to central servers while improving their effectiveness in large systems subjected to resource constraints. The ability of AI systems to constantly evolve and adapt them makes them highly suitable for complex and dynamic smart environments. Finally, the architecture for threat management improved through artificial intelligence highlights automation and response in the on-going events.

It means that in the event of a threat, the AI system can respond rapidly by isolating infected devices, closing malicious communications, or taking countermeasures on its own. DDoS attack or breach of data may be contained faster and potentially with less damage because less human intervention is needed. It is also compatible with current IoT security protocols, and it thus supports easy interaction as well as easy evolving of increased complexity within smart environments. AI-driven frameworks are made up of advanced machine learning, real-time learning, predictive analytics, and automatic reaction for building a robust, adaptive, and scalable solution to threat management through networked and ever-evolving smart ecosystems of the future (Figure 5.3).

5.5 CASE STUDIES AND APPLICATIONS

Real examples and case studies of AI-augmented threat management demonstrate how effective AI can be in protecting complex, networked smart environments. An example is the use of AI in smart cities. Here, in such cities, AI systems keep a very large volume of related data from IoT devices such as energy meters, traffic sensors, and security cameras. Through anomaly detection using machine learning techniques, AI can quickly and easily identify various unusual behaviors, such as unusual traffic patterns, unauthorized access to limited areas, or hacked devices. An artificial intelligence system was used in one case to detect a DDoS attack on a city's public transit system. This system actually limited the speed of the attack, ensuring that infrastructure functions in this urban center remained operational as it automatically isolated the compromised devices. The dynamic nature of the AI model allowed the system to learn from the incident and respond to similar incidents in the future, thereby improving the resilience of the system over time.

The healthcare sector is an important domain where artificial intelligence is used for threat management, as protecting sensitive patient information and related medical devices from cyber threats is of immense importance. Blockchain with AI can be used for correlation with ML algorithms applied to managing a patient's history and medical records [19]. Hospitals began to use AI-infused security solutions against ransomware and data leaks as well as the physical dangers of med-equipped tampering. Machine learning algorithms continuously check data generated from various sources such as equipment and patient records to detect any inconsistency that would be a possible indication of an attack. For instance, one hospital utilizing artificial intelligence was successful in identifying an interlinked diagnostic device that had been manipulated to send confidential information to other unapproved external locations. The artificial intelligence system immediately detected the anomaly and initiated counteraction protocols by deactivating the machine and alerting the security personnel. Healthcare consumers are increasingly open to sharing confidential data, necessitating organizations to establish interoperability, thereby maintaining consumer trust through demonstrated reliability, transparency, and empathy in their operations [20]. Predictive analytics apply artificial intelligence to detect possible vulnerabilities in hospital networks, allowing the management to take proactive measures before threats develop. Future research can also study the influence of robotics, AI, and big data approaches on diagnostic, maintenance, and prediction tools in healthcare supply chains [21].

Table 5.4 highlights real-world case studies where AI has been successfully applied in smart cities, healthcare, industrial IoT, and critical infrastructure.

Table 5.4 Case studies and applications of AI-augmented threat management

Sector/industry	*Application*	*Outcome*
Smart Cities	AI detects DDoS attacks targeting transportation systems.	Immediate isolation of affected systems, preventing service disruption.
Healthcare	AI-powered security to protect connected medical devices and patient data.	Detected and isolated compromised diagnostic devices, preventing data breaches.
Industrial IoT	AI secures industrial control systems (ICS) from cyberattacks.	Prevented a cyberattack from manipulating machinery, avoiding system downtime.
Critical Infrastructure	AI monitors and mitigates threats in smart grids and utilities.	Prevented cyberattacks that could have disrupted energy supply, ensuring continuity.

AI-driven threat management systems have been employed in the industrial domain for the protection of smart factories and critical infrastructures. IoT sensors and devices are strictly needed for the monitoring of machines, supply chains, and production lines in such industrial environments. With edge computing and anomaly detection, AI solutions offer real-time security monitoring in these environments. Consider, for example, the manufacturing plant, which frustrated cyberattacks on its industrial control systems by imposing AI-threat detection mechanisms. Even routine medical procedures, such as intravenous injections and blood draws, can benefit from technology, like projecting human vein maps onto the skin [22]. It is the artificial intelligence framework that detected signs of an attack aimed at changing machinery operation parameters by inspecting sensor data monitoring equipment functionality, which would eventually lead to machinery failures and disruption of the production processes. Once the threat was detected, the system automatically isolated the infected machines and responded in a manner to mitigate further damage. In this scenario, AI's continuous learning ability meant that it could identify even the most sophisticated threats. These case studies illustrate the ways through which adaptive capabilities in prediction and response in real time have made AI an indispensable tool in managing diversities of smart environments' threat architecture (Figure 5.4).

5.6 CHALLENGES AND LIMITATIONS

Although AI-assisted threat management has a number of significant benefits for the protection of intelligent environments, several issues and constraints need to be solved before such potential can be fully exploited. The

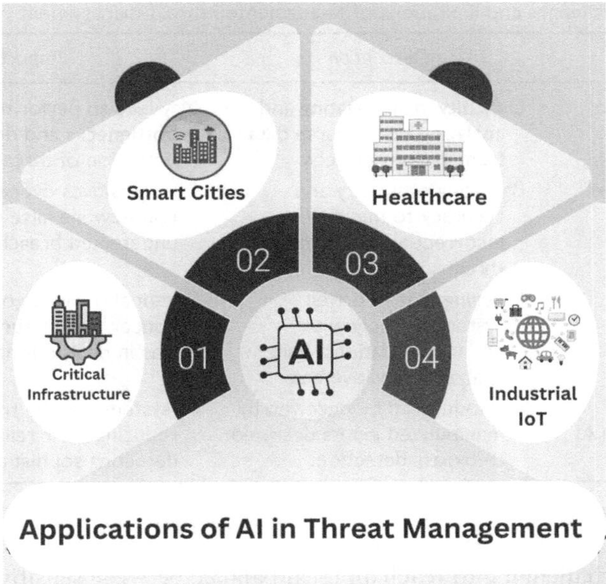

Figure 5.4 Applications of AI in threat management.

biggest challenge is in the complexity and diversity of the data generated within these ecosystems. Intelligent environments are "composed of large numbers of devices, sensors and systems, where each one produces data with dimensions, forms, and sensitivity levels." The process of efficient data aggregation and analysis depicts massive challenges, as the artificial intelligence models require the understanding of massive and diversified datasets for reliable anomaly and threat detection. In some scenarios, the volume of data can lead to complicated situations or cause a lag in the real-time detection of threats. High-quality, labelled datasets are yet another crucial factor for the effective function of AI models in supervised learning. Such deficiency, in adequate and representative training data, can make it hard for the model to identify new risks.

Table 5.5 outlines key challenges in deploying AI-powered security systems, such as data complexity, false positives/negatives, privacy concerns, and adversarial attacks.

Another critical flaw in the design of AI-based threat detection systems is the risk of false positives and false negatives. For instance, in extremely dynamic and complex environments, such as intelligent cities or industrial centers, AI may suffer from false positives as normal activities are misinterpreted as threats; therefore, security practitioners may receive unnecessary work leading to ineffective operational performance. However, because fraudsters increasingly use sophisticated methods to evade detection, false negatives – that is, actual threats missed by the system – present huge risks.

Table 5.5 Challenges and limitations of AI-augmented threat management

Challenge	*Description*	*Impact*
Complexity of Data	Difficulty in aggregating and analyzing vast, diverse data from various devices.	May lead to performance bottlenecks and delayed detection of threats.
False Positives and False Negatives	Balancing sensitivity and accuracy to minimize incorrect alerts or missed threats.	False positives overwhelm security teams, while false negatives risk undetected breaches.
Privacy and Ethical Concerns	Ensuring that AI-driven systems comply with privacy regulations while analyzing sensitive data.	Potential breaches of privacy, and concerns over surveillance and bias in decision-making.
Adversarial Attacks on AI Systems	AI models can be deceived by manipulated inputs designed to bypass detection.	AI systems can be tricked, reducing their reliability in detecting sophisticated threats.

It may be challenging to reach an ideal balance between sensitivity and precision during the optimization of AI models at the refinement stage, demanding constant re-tuning to ensure optimal operation. Another challenge involved in adversarial attacks on AI systems, where individuals with malicious intent manipulate the input data to deceive AI models into making wrong decisions, thereby making traditional defenses irrelevant. Challenges related to the scalability and adaptability of AI-augmented power system DTs were explored, along with ethical and regulatory challenges spanning data privacy, security, and trustworthiness [23]. Another major issue that reflects it is a barrier to AI-enhanced threat management: privacy and ethics concerns.

With the increasing use of AI systems in scanning vast amounts of private and sensitive data to spot threats, privacy becomes even more critical. Although methods such as differential privacy and homomorphic encryption can help alleviate some of these issues, they often do so at the cost of computing performance. The development of AI systems must also be unbiased to prevent accidental discrimination against certain user groups or undue focus on some threat vectors at the expense of others. Moreover, regional cybersecurity and data privacy standards differ and, thus far, keep changing; adherence to them would add even more complexity. These challenges call for further research, transparency, and accountability in the implementation of AI-based security control mechanisms in intelligent environments (Figure 5.5).

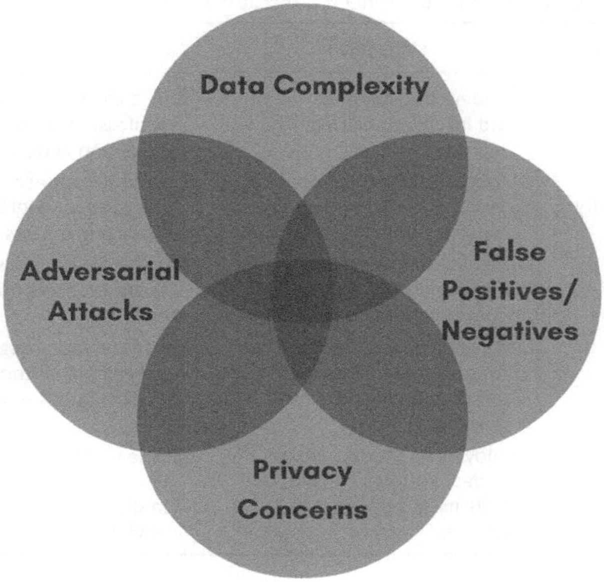

Figure 5.5 Challenges in AI-based threat detection and management.

5.7 FUTURE DIRECTIONS

Future advancements of AI-driven threat management in intelligent settings include some innovative developments promising improvement in the security, flexibility, and effectiveness of those systems. Explainable AI (XAI) is one of the most promising approaches, and work is being done to make the decision-making process by AI comprehensible and transparent. As these AI systems become widely embedded in threat detection and response – in very high-stakes environments – security professionals are constantly required to understand how the underlying AI models make decisions. XAI can prove to be of great help to instill trust in such AI systems where a human operator is able to make justifiable decisions based on insights produced by AI. Moreover, federated learning further allows decentralized training of models across multiple devices while keeping personal data not publicly accessible, hence leading to improved privacy as well as the accuracy and robustness of AI models within distributed intelligent settings. As the capabilities of AI systems improve with their ability to learn and counter new emerging threats, the second generation would result in adaptive AI frameworks capable of autonomous adaptation of new, yet unexplored attack methodologies. Without human aid, AI autonomous cybersecurity systems can identify weaknesses, analyze danger factors, and put in defenses. These will analyze the possible attack vertices by using predictive analytics and make

Table 5.6 Future directions in AI-augmented threat management

Future direction	*Description*	*Benefit*
Explainable AI (XAI)	Providing transparency and interpretability in AI decision-making.	Builds trust and allows security professionals to understand AI-driven actions.
Autonomous Cybersecurity	AI systems that autonomously detect and respond to threats without human intervention.	Reduces response time and operational overhead in managing threats.
Quantum-Resilient AI	Designing AI systems to withstand potential attacks from quantum computers.	Ensures long-term security as quantum computing evolves.
Integration with Blockchain	Leveraging blockchain for tamper-proof data exchanges and communication in AI systems.	Enhances data integrity and prevents unauthorized access in distributed environments.
Edge AI for Scalability	Deploying AI models closer to data sources to handle security at scale in large systems.	Enables real-time threat detection and response in distributed smart environments.

mitigation mechanisms proactive for them. In addition, with the advancement in quantum computing, the importance of quantum-resilient AI will be increasing. Probably the need for developing new cryptographic protocols and defense methodologies that prevent quantum systems from efficiently computing will be crucial to make the model of artificial intelligence survive the diverse challenges that quantum attacks will bring.

Table 5.6 summarizes the emerging trends and advancements in AI-driven threat management, including explainable AI, autonomous cybersecurity, and quantum-resilient AI.

The future in the immediate times is such that artificial intelligence will be seamlessly connected with the other latest technologies including digital twins, blockchain, and more. A digital counterpart or a set of digital counterparts, of an "Intelligent" or "Smart Products," have been developed to let any user or stakeholder access the attributes and services of the Smart Product during its whole life cycle [24]. Blockchain applications ensure secure, unalterable communication between AI models and Internet of Things devices in data transactions, thus further fortifying the overall integrity of the system. This will allow the AI to model, test, and predict behaviors within smart environments using models known as digital twins-a virtual replica of a real system. A blockchain can bring traceability and transparency as major benefits. It can improve information security and trust and enhance efficiency [3]. Danger detection and disaster recovery strategies would improve. Another factor with AI-driven systems is that the scalability and edge AI issues are important when the size and complexity of the smart environments grow. These systems will make it easier to manage security at scale without overloading central processing units by placing AI models closer to the edge of a network. Open radio access network (RAN) is an emerging framework for network transfer through infrastructure virtualization and embedded intelligence to provide end users with more stable network connectivity services and advanced capabilities [25]. This way, big dispersed environments will support decisions in real time. This will be particularly crucial to the growth of smart cities and the industrial Internet of Things applications. The point is that as these technologies advance, artificial intelligence will be there to develop secure, dependable, and flexible intelligent environments (Figure 5.6).

5.8 CONCLUSION

In summary, AI threat management will deeply change the security framework in emerging smart environments since it provides sophisticated, flexible, and scalable solutions to handle the complexities of modern cyber and physical threats. As smart environments – ranging from smart cities to industrial Internet of Things systems – become more networked and

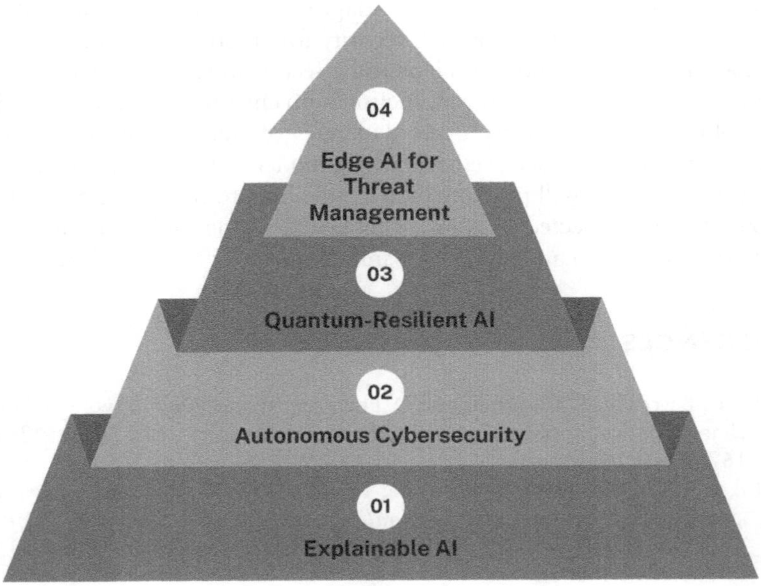

Figure 5.6 Future of AI – augmented threat management.

data-driven, the capabilities offered by artificial intelligence in processing massive amounts of real-time data using machine learning, anomaly detection, and predictive analytics enable identifying hazards and mitigating them before they escalate. AI-powered solutions fusing automation and real-time response help speed up the process of even more effective countermeasures, securing critical infrastructures, personal data, and operational integrity in these dynamic, interconnected ecosystems. Block IoT Intelligence claims the mitigation of existing challenges to obtain high accuracy, reasonable latency, and security [26].

Despite its promises, several challenges that need to be addressed to enhance the effectiveness of AI-led security measures include data complexity management, balance between false positives and false negatives, and ethical and privacy-related resolution regarding the wide integration of AI technologies. More than that, AI models have to show resilience against malicious attacks as they attempt to adapt and improve continuously to remain on par with the ever-evolving threat environment. It requires proper federated learning schemes, advanced integration of blockchain and digital twins, and continuous investigation and advancement in explainable AI to ensure that AI systems remain private, integrity-guaranteed, and adaptable. Its future development in the field of threat management would depend upon how it would evolve to better manage increasingly complex threats, learn adaptation to emerging threat vectors, and scale appropriately for ever-growing complexity.

Combining quantum-resilient AI, adaptive AI frameworks, and state-of-the-art techniques such as edge AI, security for smart environments will take the lead in being even more resilient and effective in real time. The intelligent use of artificial intelligence in conjunction with emerging technologies is likely to revolutionize the security game dynamics of smart environments into a more proactive, responsive, self-sustaining model. Artificial intelligence-based threat management will fundamentally contribute to the secure and safe evolution of interconnected systems, thus safeguarding end-users and critical infrastructure within an increasingly interconnected global landscape.

REFERENCES

[1] B.R. Barricelli, E. Casiraghi and D. Fogli A survey on digital twin: Definitions, characteristics, applications, and design implications, *IEEE Access* 2019, 7, 167653–167671.

[2] Aryan Kaushik, Rohit Singh, Ming Li, Honghao Luo, Shalanika Dayarathna and Rajitha Senanayake, *Integrated Sensing and Communications for IoT: Synergies with Key 6G Technology Enablers*, 2024, IEEE, 10.1109/IOTM.001.2400052

[3] Meng-Leong How and Sin-Mei Cheah, *Forging the Future: Strategic Approaches to Quantum AI Integration for Industry Transformation*, MDPI, doi.org/10.3390/ai5010015

[4] C. Mistry, U. Thakker, R. Gupta, M. S. Obaidat, S. Tanwar, N. Kumar and J. J. P. C. Rodrigues, "MedBlock: An AI-enabled and blockchain driven medical healthcare system for COVID-19," in *Proc. ICC IEEE Int. Conf. Commun.*, Jun. 2021, pp. 1–6, doi:10.1109/ICC42927.2021.9500397

[5] Eryk Schiller, Elfat Esati and Burkhard Stiller, *IoT-Based Access Management Supported by AI and Blockchains*, MDPI, doi: 10.3390/electronics11182971

[6] IIoT: Le Guide Complet Pour Bien Lancer Votre Projet, Ozone Connect, Toulouse. 2022. Available online: https://iotindustriel.com/actualites-et-evenements/ozone-connect-lance-le-1er-guide-iot-industriel-au-grand-maghreb/ (accessed on 6 May 2022).

[7] Mohamed Abdel-Basset, Rehab Mohamed and Victor Chang, *A Multi-Criteria Decision-Making Framework to Evaluate the Impact of Industry 5.0 Technologies: Case Study, Lessons Learned, Challenges and Future Directions*, SPRINGER, doi.org/10.1007/s10796-024-10472-3

[8] Ama Ranawaka, Damminda Alahakoon, Yuan Sun and Kushan Hewapathirana, *Leveraging the Synergy of Digital Twins and Artificial Intelligence for Sustainable Power Grids A Scoping Review*, MDPI, doi: 10.3390/en17215342

[9] R. Singh and S.S. Gill Edge AI: A survey. *Internet Things Cyber-Phys. Syst.* 2023, 3, 71–92.

[10] B. McMahan, E. Moore, D. Ramage, S. Hampson, y B.A. Arcas Communication-efficient learning of deep networks from decentralized data. In *Proceedings of the 20th International Conference on Artificial Intelligence and Statistics, Fort Lauderdale*, FL, USA, 20–22 April 2017; pp. 1273–1282.

[11] C. Kishor Kumar Reddy, P.R. Anisha, B. Tirupathi Reddy, Rambabu D. Srinivasulu. LightWeight RealTime Weather Forecasting Simulation Over Bangladesh Using Deep Learning. *INTJECSE*, vol. 449465, 2022, 4616–4633.

[12] Iqbal H. Sarker, *AI-Based Modeling: Techniques, Applications and Research Issues Towards Automation, Intelligent and Smart Systems*, SPRINGER, doi: 10.1007/s42979022-01043-x

[13] Leila Ismail and Rajkumar Buyya, *Artificial Intelligence Applications and Self-Learning 6G Networks for Smart Cities Digital Ecosystems: Taxonomy, Challenges, and Future Directions*, MDPI, doi: 10.3390/s22155750

[14] Radosław Wolniak and Kinga Stecuła, *Artificial Intelligence in Smart Cities Applications, Barriers, and Future Directions: A Review*, MDPI, doi: 10.3390/smartcities7030057

[15] Abdulkhaliq Ali F. Alshadidi, Ahid Amer Alshahrani, Lujain Ibrahim N. Aldosari, Saurabh Chaturvedi, Ravinder S. Saini 1, Saeed Awod Bin Hassan, Marco Cicci and Giuseppe Minervini, *Investigation on the Application of Artificial Intelligence in Prosthodontics*, MDPI, doi: 10.3390/app13085004

[16] Alexander Blanchard andMariarosaria Taddeo, The Ethics of Artificial Intelligence for Intelligence Analysis: a Review of the Key Challenges with Recommendations, Springer, doi: 10.1007/s44206-023-00036-4

[17] Guneet Kaur Walia, Mohit Kumar and Sukhpal Singh Gill, *AI-Empowered Fog/Edge Resource Management for IoT Applications: A Comprehensive Review, Research Challenges and Future Perspectives*, IEEE, doi: 10.1109/COMST.2023.3338015

[18] Hoang-Sy Nguyen and Miroslav Voznaka, *Bibliometric Analysis of Technology in Digital Health: Exploring Health Metaverse and Visualizing Emerging Healthcare Management Trends*, IEEE, doi: 10.1109/ACCESS.2024.3363165

[19] Martin Barton, Roman Budjac and Pavol Tanuska, *Gabriel Gaspar and Peter Schreiber Identification Overview of Industry 4.0 Essential Attributes and Resource-Limited Embedded Artificial-Intelligence-of-Things Devices for Small and Medium-Sized Enterprises*, MDPI, doi: 10.3390/app12115672
[20] K. Wolitzky, R. Fivush, E. Zimand, L. Hodges and B. O. Rothbaum," Effectiveness of virtual reality distraction during a painful medical procedure in pediatric oncology patients," *Psychol. Health*, vol. 20, no. 6, pp. 817–824, Dec. 2005, doi: 10.1080/14768320500143339
[21] Nasif Fahmid Prangon and Jie Wu, *AI and Computing Horizons: Cloud and Edge in the Modern Era*, MDPI, doi: 10.3390/jsan13040044
[22] K. B. Letaief, Shi Y. Chenw, et al. The Roadap To 6G: AI Empowered Wireless Networks[J]. *IEEE Communications Magazine*, 2019, 57(8): 84–90.
[23] Vasile-Daniel Pavaloaia and Sabina-Cristiana Necula, *Artificial Intelligence as a Disruptive Technolog A Systematic Literature Review*, MDPI, doi: 10.3390/electronics12051102
[24] A. AI. Yarali, 5G, and IoT. In *Intelligent Connectivity: AI, IoT, and 5G; IEEE*: London, UK, 2022; pp. 117–131, ISBN 978-1-119-68523-4.
[25] Nadine Y. Fares, Denis Nedeljkovic, and Manar Jammal, *AI-enabled IoT Applications: Towards a Transparent Governance Framework*, IEEE, 10.1109/GCAIoT61060.2023.10385106
[26] Nouf M. Alzahrani and Faisal Abdulaziz Alfouzan, *Augmented Reality (AR) and Cyber Security for Smart Cities—A Systematic Literature Review*, MDPI, doi: 10.3390/s22072792

Chapter 6

Network security governance framework for cloud-integrated IoT systems

Sanjay Poddar, Ram Chandra Sachan, Mariyam Ouaissa, and Mariya Ouaissa

6.1 INTRODUCTION

Today, the merging of cloud computing into networks and integration with IoT has reshaped modern landscapes into smarter, more seamless Artificial Intelligence of Things (AIoT) [1]. The synergy enables real-time data mining, risk-free device communication, and data computation as needed over multiple applications such as smart city, industrial IoT, and healthcare [2]. Yet, these developments also increase the demand for strong network security governance to secure sensitive information, as well as overall management and organization of those larger-scale, integrated IoT networks that exist in the cloud. With Internet of Things (IoT) devices being deployed in millions, securing them and their communication networks is an even harder problem. When considering the fundamental differences with IoT—many devices do not have processing capability (it is small, low power and widely spread, often in uncontrolled environments)—it quickly becomes clear that traditional network security measures will never meet this challenge. The incorporation of cloud services into IoT networks brings added complexity which introduces challenges for privacy, unauthorized access, and abuse of cloud resources [3].

With the large-scale adoption of cloud-integrated IoT systems, network security governance has emerged as a vital research area where future direction is needed due to the unique threats in AIoT environments and also for improving upon current frameworks that have been established for other domains such as Information Technology (IT)-based systems and/or legacy systems which do not involve cloud-integrated environments [4].

This chapter explores challenges in traditional network security governance including those faced by existing service-oriented computational models when applied to AIoT. While reviewing adaptation proposals of popular recommendations like NIST 800-53, we highlight areas where even more skills are required for establishing effective frameworks suited especially in these evolving premises. In this chapter, we will discuss the architecture of IoT systems that integrate with cloud and some vulnerabilities at both ends along with potential AI-based solutions for securing networked

DOI: 10.1201/9781003606307-6

IoT ecosystems. We will also discuss the standards of practice and suggest a governance framework that would help secure these networks and cover their continuous monitoring, secure communication, and compliance with industrial standards.

6.2 CLOUD-INTEGRATED IoT SYSTEMS

Recent rapid and widespread usage of cloud computing technology has proven to be the game changer for handling IoT systems by providing non-stop scale, storage and real-time data ingestion [5]. Traditional IoT managed devices in closed networks and local or edge data processing. With cloud integration, however, IoT devices can join a centralized platform where data can be stored and analyzed over large networks with minimal delay as possible. While this architectural transition has certain operational benefits, it also poses new security governance challenges.

6.2.1 Cloud-integrated IoT system architectures and components

A cloud-integrated IoT system is made up of loosely coupled components where each component has its role to play in the functionality and security of the IoT ecosystem [6]. The key components include:

IoT devices: These endpoints in the network gather and send data. They cover a very broad range, such as sensors in smart cities and factories to wearable devices in healthcare. IoT devices are often low-powered and may not have strong security built-in, making them a target for attacks.

Edge gateway: Edge gateways are the intermediary devices between IoT and cloud. These are vital as gateways for combining data from multiple devices, pre-processing of that data, and quick latency by filtering out unrequired data before sending it to the cloud. The in-vehicle edge gateways also act as a security checkpoint where they often count the encryption and device authentication functionality to protect the data going in and out of the cloud.

Cloud forwarding: The cloud platform is where we store all IoT data. IoT leverages cloud computing services ideal for the large-scale data analysis required to process massive amounts of IoT information in real time, enabling predictive maintenance, anomaly detection, and AI-driven decisions. But the cloud also brings its own set of security challenges including unauthorized access, data breaches, and compliance with various data protection rules and regulations.

Application layer: The top layer is responsible for the software applications and user interfaces via which users interact with the IoT devices and cloud data. These applications can be as simple as a mobile app for individual users or complex such as industrial control systems and are responsible for monitoring and controlling IoT networks. It is important to keep the access and communication between application layer and cloud secure as if they are compromised, then the unauthorized entry to the server becomes easy leading to leakage of data.

6.2.2 The role of cloud in expanding IoT capabilities

The integration of devices with the cloud has drastically increased what IoT systems can do to the point where they are scalable and have seamless interconnectivity across various locations and devices. Through cloud platforms, companies are able to deploy IoT networks which can be scaled at large without the need for extensive on-premises infrastructure. The cloud offers on-demand computation power, allows processing and analyzing data in real-time when otherwise impossible to achieve with single IoT devices or edge gateways. Additionally, cloud platforms provide features of AI and ML tools that can further develop IoT applications. For example, cloud-based AI solutions can use IoT sensor data to predict when equipment will fail or optimize energy consumption or analyze network traffic for anomalies. This feature is particularly useful in industries like manufacturing, healthcare, and smart cities since data-based insight can enhance operational efficiency and safety [7].

6.2.3 Security aspects of cloud-enabled IoT networks

While cloud integration is beneficial, it poses several security issues that need proper governance too. The cloud-native characteristic of IT networks brings up various specific threats to IoT networks due to their distributed, large-scale nature and dependency on third-party cloud providers [8]. Here are some important security considerations:

Compliance and data privacy: Sensitive data is transmitted in the cloud and thus requires being compliant with many data protection regulations like GDPR or HIPAA. Reputational harm and financial penalties due to breaches of data privacy necessitate that organizations implement strong data encryption, access controls and audit mechanisms.

Device authentication and identity management: In an IoT system with cloud integration, authenticating each connected device to the network is essential to allowing only authorized devices. Identity management solutions, such as Public Key Infrastructure (PKI), are essential for creating trusted identities for devices, applications, and users on the IoT.

Network segmentation and access management: Once a network breach happens, the attacker movements from one host to another will begin.

There is an option possible for that too, so ensure you have your network segmented and remove unintended lateral movement paths. Segmentation provides the ability to isolate IoT devices, cloud applications, and data stores from each other, allowing fine-grained access control. This minimizes the impact of a compromised device or application, reducing the risk of widespread network disruption.

Threat detection and incident response: It is difficult to detect threats in real-time especially when the same environment is shared by multiple clouds such as third-party communication, virtual networks, alliances, etc., rightly like we have for cloud integrated IoT systems. Using Security Information and Event Management (SIEM) solutions along with AI-powered analytics, security professionals can continuously monitor network activity and receive alerts for any anomalous actions. This enables them to respond proactively to incidents, mitigating the damage from cyberattacks.

Cloud security risks: Organizations need to be aware of all the security risks involved in the shared responsibilities model when adopting cloud services. The security of the infrastructure is managed by cloud providers, and their customers manage their data and applications within the cloud. Awareness about these responsibilities is vital because overlooking them can lead to misconfigurations, data leakages, and vulnerabilities in the cloud environment.

With enterprises implementing more and more complex AIoT systems, governance requires a deep understanding of the architecture and security aspects of cloud-based IoT networks. In the subsequent sections, we will discuss frameworks and best practices that can guide us in securing these systems and establish a comprehensive strategy for cyber defense governance of network security assets for IoT environments connected to cloud.

6.3 NETWORK SECURITY GOVERNANCE FRAMEWORKS FOR AIoT

With IoT systems integrated in the cloud becoming increasingly large and complex, a governance framework should be designed to take into account security needs specific to these systems. Governance frameworks offer a structured set of guidelines and best practices for ensuring that the security measures taken (or added) to an organization align with its objectives, regulatory requirements, and industry standards. Such governance takes on more complexity in AIoT environments as it must cover the span of device security, data privacy, and access control to ensure real-time monitoring for both the IoT-enabled devices and the cloud platform [9].

Governance frameworks like ISO and NIST are some of the examples that we will cover in this section, which may facilitate their adaptation for network security in AIoT systems. This post elaborates on the applicability of these frameworks to cloud-enabled IoT systems and describes how they should be customized to address the requirements of AI-based cloud-integrated smart environments [10].

6.3.1 Review of major governance frameworks

ISO/IEC 27001—Information Security Management Systems (ISMS): ISO/IEC 27001 is a broad range standard that helps you in managing information security across your organization. Guidelines on risk assessment, asset management, access control, incident response, and compliance monitoring are included. ISO 27001 helps organizations design an ISMS for integrated IoT systems with the cloud by addressing risks associated with interconnectedness of devices, data uploads to the cloud, and real-time transmission of data. ISO 27001 implementation not only enables protection for information in an AIoT ecosystem, but can help mitigate weaknesses and foster an ethos of ongoing improvement of systems security.

NIST Cybersecurity Framework (CSF): The five core functions of the NIST CSF—Identify, Protect, Detect, Respond and Recover—are renown. This framework is flexible and can help organizations to improve the governance of network security in the AIoT system. For instance, the "Identify" function focuses on managing assets and risks, which is critical for identifying IoT devices on different parts of the network. Additionally, NIST CSF also enhances capabilities in advanced threat detection response and resilience which are essential to securing cloud-connected IoT networks from cyber threats.

ISO/IEC 27017—Information Security Controls for the Cloud: ISO 27017 brings the principles of ISO 27001 to the cloud with a comprehensive list of security controls designed specifically for deploying in and using cloud environments. ISO 27017 offers advice for cloud-based data storage, managing service providers, and incident response protocols that are useful to organizations using an integrated cloud with their IoT systems. This framework is particularly useful for cloud-based IoT networks as it relates to protecting data, controlling user access and encryption on cloud-based data in IoT.

6.3.2 Zero-Trust Architecture (ZTA)

Zero-Trust is a relatively new security model based on never trust, always verify. In fact, in a zero-trust model, all network access requests are authenticated, authorized, and continuously validated, whether from outside or inside the organization. Zero-trust for AIoT systems means using access control policies and device authentication, with segmented networks that restrict the impact of breaches. Zero-trust proves especially useful when applied to cloud-integrated IoT environments, where both remote access and communication between devices are commonplace.

6.3.3 CIS controls

The CIS controls are a set of best practices developed by the Center for Internet Security (CIS) that aim to protect against widespread cyber threats. Controls reviewed range from hardware inventory, software asset management (SAM), data protection to monitoring over a network. The CIS controls are directly applicable to cloud-integrated IoT networks and detail the recommendations for securely protecting network devices, monitoring unauthorized access or exploitation of organizational systems and services, and monitoring the use of secure configurations for hardware and software.

6.4 THREAT LANDSCAPE AND VULNERABILITIES OF CLOUD-INTEGRATED IoT NETWORKS

While these two giants of technology, the cloud and the IoT, made us more connected and functional than ever, they also opened a whole new landscape for security vulnerability. The open and complex interactions between distributed environments, along with the variety of devices included within an IoT network, add to its total attack surface, meaning that a cloud-integrated IOT also represents an extremely high vulnerability. Knowledge of the unique threats and vulnerabilities that endanger these systems is essential for sound governance of network security.

In this section, we will explore the key threats and vulnerabilities that cloud-integrated IoT networks might encounter, along with real-world examples to illustrate what security breach can lead to. Making sense of these risks enables organizations to formulate tailored governance strategies to reduce fragilities across their AIoT ecosystems [11].

6.4.1 Common threats and vulnerabilities

Device compromise and unauthorized access: Most IoT devices are designed to minimize processing power and memory, with the result that they often lack strong security. Attacker targets these devices due to weak authentication because they ship with default credentials. An example of this can be found in compromised devices that serve as entry points into the network—to enable attackers to perform lateral movement and reach more sensitive resources elsewhere in the cloud infrastructure.

Data privacy and integrity: IoT has various data, and this can be sensitive; it may include user details, operational details, and infrastructure information. The data you transmitted and saved in the cloud can be intercepted, altered, or accessed by another party. The consequences of data breaches can be dire indeed, such as exposure of PII (personally identifiable information), violation of data protection laws and regulations and loss of trust from users and stakeholders alike.

Distributed Denial-of-Service (DDoS) attacks: Specifically, IoT devices are one of the most common types of devices to be turned into bots so that attackers can leverage them for Distributed Denial-of-Service (DDoS) attacks. Attackers can overwhelm cloud-integrated IoT systems with massive amounts of traffic from infected devices by flooding the network, ultimately destabilizing these systems and causing service interruptions rendering the system severally compromised. Such issues are particularly worrying in mission-critical devices like those used in the healthcare and industrial IoT, where downtime can be damaging.

Firmware and software vulnerabilities: IoT devices can be some of the worst offenders, with many operating on outdated firmware and software that might easily contain vulnerabilities for attackers to exploit. In contrast with traditional IT systems where automated updates are common, IoT devices may remain exposed to known vulnerabilities for an extended period. In IoT networks, since they are integrated with cloud, a single vulnerable device can be a weak chain which leads to attacks on an entire network.

Insider threats: Cloud-connected IoT environments are further complicated by the number of different parties with access, from employees to contractors and third-party vendors, making them a continuing source of risk from insider threats. Insider threats can stem from either malicious or unintentional acts, including configuration issues. An insider could easily take advantage of such access to either compromise network security or leak sensitive data if there is no strict control and monitoring.

Man-in-the-Middle (MITM) attacks: Due to the communication of IoT devices, over unencrypted or poorly secured channels, they are vulnerable to Man-in-the-Middle (MITM) attacks. In the case of a MITM attack, the attacker intercepts communication between the IoT device and cloud platform (and who knows what else). This may result in data alteration, unauthorized entry into the IT system and command parameters manipulation, which are all detrimental to the trustworthiness of the IoT network.

6.4.2 Real-world security breaches

Mirai botnet attack: Perhaps the most notorious case of mass IoT hacking is the Mirai botnet. Mirai had leveraged passwords and exposure to the internet implementations of IoT devices to spread across thousands in 2016. These compromised devices were subsequently deployed to execute a giant DDoS blitz against target sites and services. It drew attention to many weaknesses in IoT devices that are exploitable for botnets, and it led the crypto community to also consider stringent access controls and authentication between remote connected devices.

Smart home network data breach: A breached smart home IoT network showed that memorizing attack paths across smart thermostat and security camera systems allow unauthorized access to user data stored in the cloud. It allowed the attackers to not only capture unencrypted communication between devices and the cloud but also manipulate settings and log in to user accounts. WB-13 what went wrong: Users should have trusted the encryption (which this breach never gave them) and not performed the traffic in plain text leading to widespread snooping into the data stream. This pointed out a gap in cloud-integrated IoT environments as mainly secure communication channels were a weakness which prevented user from maintaining privacy.

Industrial IoT attack at manufacturing facilities: In this attack, the cloud-integrated IoT network of a manufacturing facility was targeted, wherein IoT sensors and controllers were deployed to monitor and manage production processes. The facility had IoT devices with unfixed firmware, which attackers exploited to take control of important machinery. Attacker interrupted the production and then asked a ransom to restore it. The take-home from the case was that IoT devices with outdated firmware can be a vulnerability, and any incident on these devices can have an impact on industrial environments.Taking a closer look at the risks linked with cloud adoption, the combination of IoT with specific cloud platforms creates some risks that are not necessarily covered by traditional network security frameworks. The complexity increases with cloud integration as the organizations have to secure the IoT along with the cloud which has different complexities and attack vectors.

6.4.3 Cloud-specific vulnerabilities

Cloud providers implement a shared responsibility model, where they protect the infrastructure, and customers protect their data, applications, and devices. General misunderstandings surrounding these responsibilities can create opportunity gaps for security—especially in the context of IoT applications that leverage cloud integration. One common potential pitfall is complete reliance on cloud providers for security. If customers believe that they do not need to take any extra effort toward protecting their data in the cloud, they might skip critical measures such as encryption, access control, and monitoring.

API vulnerabilities: The cloud-integrated IoT systems depend on APIs to facilitate the communication between IoT devices, edge gateways, and cloud services. When these APIs are not protected and secured accurately, attackers can exploit them to gain access to cloud resources and perform unauthorized actions like modifying data or even compromising the operations. In IoT networks, which are highly integrated with the cloud and often have widely exposed API endpoints, APIs become a significant source of vulnerabilities that are never monitored at that level.

Data exfiltration risks: The transfer of sensitive data or information from the IoT network by attackers is a very significant risk in cloud-integrated IoT environments. The resale of large amounts of sensitive data is possible since IoT devices constantly send data to the cloud, and any compromise in this pipe proves that too much valuable information can get into the wrong hands. Improperly configured cloud storage or insecure networks can easily be exploited by attackers to breach data with implications on a larger scale.

Misconfigurations and shared responsibility errors: While cloud platforms provide flexibility and scalability, in IoT setup there may be some misconfigurations that make these networks susceptible to threats. Simple things like leaving a cloud storage bucket public or using weak or mismanaged credentials for the various cloud services can give attackers an access point. Without the proper configuration management and a best practices approach, one can have security holes in cloud-connected IoT environments.

6.4.4 Vulnerabilities in governance making sense of security

The security environment created by the unique characteristics of IoT devices connected to a cloud highlights the importance of an anticipatory governance framework focused on device access, data privacy, and specific risks related to the cloud. Governance strategies should consider:

Enforcing strong authentication and access controls: By requiring Multi-Factor Authentication (MFA) for device access and ensuring that only authorized users and devices gain entry to the network, organizations can decrease the risk of unauthorized access into their networks.

Encrypting data in transit: Encrypting data in transit and ensuring that all communication between IoT devices, edge gateways, and cloud platforms is secure helps prevent MITM attacks and data interception.

Routine firmware and software updates: Policies for updating and patching IoT devices in a timely manner can limit the ability of attackers to exploit software vulnerabilities on these systems due to outdated software, making it more challenging for them to follow through with an attack.

Continuous monitoring and response: Real-time monitoring enabled by SIEM or other solutions can help detect early anomalies that when left ignored can grow to become large breaches but when detected timely with relevant type of analysis work like wonders.

Industry standards compliance: Compliance with ISO 27001 and ISO 27017 demonstrates a systematic approach to security governance and helps cloud-integrated IoT systems meet security requirements.

The following section will discuss AI-enabled network security to improve threat recognition, anomaly detection, and incident response within cloud-integrated IoT environments by supplying organizations with powerful systems to protect their AIoT ecosystem.

6.5 AI-DRIVEN NETWORK SECURITY SOLUTIONS FOR IoT ECOSYSTEMS

In cloud-based IoT ecosystems, the traditional security approaches may be insufficient owing to the sheer scale, heterogeneity, and complexity in terms of devices and flows. AI provides effective features that can improve security for such networks, allowing organizations to perform real-time threat detection and response as well as prevention. IoT networks can benefit from AI-driven solutions to enable adaptive, automated, and intelligent network security strategies for IoT environments [12]. This section analyzes the role of artificial in network security for IoT ecosystems, focusing on main applications like intrusion detection, anomaly detection, and predictive threat intelligence. Finally, we will examine the advantages and disadvantages of implementing AI-powered security solutions and share best practice guidelines for incorporating these tools into engineering stewardship governance frameworks in the context of AIoT.

6.5.1 Why AI is crucial for IoT security

Real-time threat detection and response: In the large-scale IoT networks where billions of devices are connected, data is generated in petabytes every minute and receiving a warning and acting on it in real-time analysis is often quite hard with traditional methods. AI-powered systems are capable of processing huge volumes of data in minimal time, detecting any threats and responding automatically. AI algorithms, for instance, can identify patterns that are suggestive of cyberattacks, and this could include a spike in traffic or an unauthorized access attempt and immediately containment and remediation can take place.

Detection of anomalies and analysis of behavior: Different types of IoT devices have their specific operational patterns based on the nature of functions it serves, location at which they are being used, and their frequency, respectively. In this way, AI-driven systems can learn over time to recognize normal user behavior and identify deviations that could be indicative of a malicious activity. AI-enhanced algorithms can raise an alarm for security teams when an IoT device starts sending data out of its normal time windows or tries to connect to some anomaly endpoints.

Threat Intelligence, but Predictive AI, allows for predictive analytics by processing large volumes of historical data to identify potential threat vectors in the future. Machine learning models that utilize past incidents, threat intelligence feeds, and behavioral data can forecast the attack surface on potential exploitable vulnerabilities. In cloud-integrated IoT environments, it can prove highly useful to shift from a completely reactive mode of threat mitigation to a predictive approach, thereby taking steps to secure devices and critical network segments even before an attacker launches a successful assault against the digital assets that you seek to protect [13].

Automated incident response: Due to the volume and velocity of all security events, manual incident response is often not feasible in cloud-integrated IoT ecosystems. Using AI, organizations can automate isolated incidents response processes such as isolating a compromised device, blocking a malicious IP address, and running remedial actions. It speeds up the response time to limit damage to minimize work on the security team.

Data protection and compliance: AI security tools can be implemented to inspect data flows and ensure compliance with privacy legislation. For instance, it can analyze data access patterns to identify inappropriate data handling practices and prevent sensitive information from getting compromised while helping meet compliance requirements. It is greatly applicable to industries that work under guided data protection, for example, the healthcare and finance domain.

6.5.2 AI-driven tools

In this sub-section, we will discuss the AI-driven tools available for securing IoT Network.

ML-based Intrusion Detection System (IDS): Current Intrusion Detection Systems (IDS) are centered around analyzing network traffic for any form of malicious activity by utilizing various machine learning models. They apply supervised and unsupervised learning to detect the patterns of attacks like brute-force attempts, malware infections, and lateral movement on networks. In case of IoT ecosystems, IDS can serve as an additional layer of security against intrusion, monitoring communication channels between devices, edge gateways, and cloud services.

The root of this is Behavioral Analytics and User and Entity Behavior Analytics (UEBA). As a UEBA you can use AI to analyze the behavior of users, devices, and entities in your network. These systems also set a baseline of normal behavior, allowing them to spot anomalies that can signal insider threats, compromised devices, and external attacks. UEBA, for example, can flag activity as suspicious and trigger investigation if an IoT device is accessed from an unusual place or outside of normal business hours.

AI integration in Threat Intelligence Platforms (TIPs): Threat Intelligence Platforms (TIPs) collate information from various sources like threat intelligence feeds, logs, and security events. These algorithms are driven by AI to process such data and identify emerging threats while also generating risk scores which enable security teams to prioritize response actions. AI-empowered TIPs enable enterprises to proactively respond to threats relevant to an enterprise IoT ecosystem and protect themselves from attackers in advance.

NLP-based threat intelligence analysis: NLP algorithms allow artificial intelligence tools to parse unstructured threat intel data, such as security reports, news articles, and forums where hackers might post about new vulnerabilities. From this data, AI systems can track trends and threats that are relevant to an IoT network. This enables organizations to proactively manage their risk by providing up-to-date threat intelligence on how and when an attack is harder to mitigate than others.

Automated Security Orchestration, Automation, and Response (SOAR) platforms: SOAR combines smart decision-making with automatic security actions to respond faster. SOAR platforms can orchestrate responses, including isolating compromised devices, revoking access privileges, or alerting stakeholders in the context of cloud-integrated IoT networks. SOAR platforms with embedded AI can automatically respond in a context-appropriate manner to the unique features of each threat, making IoT ecosystems more resilient [14].

6.6 BEST PRACTICES FOR AIoT NETWORK SECURITY GOVERNANCE

Key components of effective network security governance for service-based internet and cloud-integrated IoT (AIoT) will include secure architecture design, monitoring, access management, and corporate standards compliance. This strategy includes the following best practices that represent a complete framework of how to make sure that it addresses network security while handling the unique requirements of AIoT ecosystems in view of any weaknesses, making them more resilient against cyberattacks. This section discusses the best practices organizations can follow in order to put a strong governance framework in place that will protect their AIoT networks and data [15].

6.6.1 Architectural security

A well-designed security architecture is the keystone of any governance of perimeter security and its defense-in-depth illustration. Security by design means planning, building, and implementing the network in such a way that security controls are included rather than bolted on after the fact. Key elements include:

Network segmentation: Divide network traffic into segments related to device type, role, and risk level. Segmentation of network devices reduces the lateral movement behavior by an attacker through the network and hence allows organizations to contain security incidents and avoid a single point of compromised device from affecting other devices in the same network.

Zero-trust model: Implement a zero-trust model based on where each device and the user must be authenticated, authorized, and continually validated. This is beneficial in AIoT networks with a large number of endpoints and heterogeneous devices, enabling fine-grained access control for distributed systems.

Data encryption: Use encryption for your data in all states; this includes the encryption of data at rest, during transmission as well as while being used. End-to-end encryption safeguards sensitive data from being accessed by anyone other than the intended recipient, even if it is intercepted by an attacker.

6.6.2 Continuous monitoring and threat detection

Effective monitoring of AIoT automatically and continuously is necessary to ensure that the company can always be aware of what is happening in this environment [16]. This gives organizations the ability to detect anomalies at an early stage and even address security incidents before they become major problems through real-time monitoring. Best practices include:

Adoption of SIEM and AI-driven analytics: Logs from IoT devices, cloud platforms, and network components can be brought together in centralized monitoring solutions like a SIEM system. By leveraging AI-guided analytics along with SIEM, organizations can process large sets of data to identify abnormal behaviors and swiftly send alerts for potential threats.

Behavior-based anomaly detection: Using models that learn the patterns of behavior in an organization, abnormal activities among devices can be identified, which may indicate malicious activity. Because devices in AIoT networks can behave differently and traditional rule-based detection can be difficult to implement, anomaly detection becomes especially useful.

6.6.3 Identity and Access Management (IAM)

Identity and Access Management (IAM) is essential for regulating access to AIOT networks and resources. Considering the magnitude of IoT Networks, access to sensitive data and functions should be restricted to only authorized entities. Best practices include:

Multi-Factor Authentication (MFA): MFA should be applied to all devices and user accounts—MFA provides another level of security beyond password-based logins. It protects against unauthorized access, even with stolen credentials.

Role-Based Access Control (RBAC): Set access rights according to user roles and restrict users from accessing more resources than needed in order for them to perform their function. This reduces the impact of insider threats and also limits the damage an attacker could cause if they manage to break into your system.

Device authentication and certificate-based trust models: Leverage digital certificates and Public Key Infrastructure (PKI) to authenticate devices so that only trusted devices can connect as members of a network. This method of trust modeling based on certificates serves well especially for large-scale IoT networks, which typically consist of many devices and tend to be highly remote.

6.6.4 Incident response and recovery

In an AIoT environment, downtime or data loss due to a lack of incident response plan can have dire consequences, making it extremely important for organizations to clearly define their incident response plans [17]. Best practices for incident response include:

Automated response playbooks: Create playbooks that define automated responses to common types of security incidents, including a compromised device or an unauthorized access attempt. Utilizing automated playbooks will minimize the response times and be much more efficient in consistency for handling an incident.

Regular backups and recovery mechanisms: Repeat backup of important data and define processes to restore date and services after an incident. Redundancy and recovery protocols in case of provocation such as network failure or security breach to ensure that the operation continues.

Threat simulation: Periodic threat simulations, such as red-teaming exercises to assess the efficacy of the incident response plan. They help identify gaps in security posture and prepare the teams for real-life incidents.

6.6.5 Regulatory compliance

Compliance with industry standards and regulatory frameworks offers a systematic governance framework for network security, which addresses the need to ensure that AIoT environments meet predefined security metrics [18]. Best practices to consider for compliance include the following:

Adopting pertinent standards (ISO 27001, NIST CSF): Align security governance with common standards such as ISO 27001 and the NIST Cybersecurity Framework. They provide clear framework for responsible security risk management and have become the gold standard for securing data, protecting access to your systems, and responding in an incident.

Frequent compliance audits: Execute regular audits for monitoring adherence with standards of the industry and systematic policies. Audits identify areas of non-compliance, contributing toward improving the overall network security practices.

Documentation of policies and staff training: All security policies, procedures, and protocols should be well documented, and staff should receive regular training. That promotes psychological safety and a culture of responsibility around security in an AIoT world.

6.7 CONCLUSION

This chapter has presented a comprehensive Network Security Governance Framework for cloud-integrated IoT systems, addressing the multifaceted security challenges posed by their dynamic and distributed nature. The framework integrates governance principles, automated threat detection, and policy enforcement to ensure a holistic approach to securing IoT ecosystems. Through the proposed governance model and risk assessment strategies, organizations can effectively manage vulnerabilities and maintain compliance with regulatory standards. The case study on a smart healthcare system highlighted the framework's practical applicability and effectiveness in mitigating security threats while maintaining system scalability and performance. This framework serves as a critical step toward enhancing the resilience and trustworthiness of cloud-integrated IoT systems, paving the way for secure adoption in various critical domains. Future work will focus on expanding the framework to incorporate emerging technologies such as AI-driven anomaly detection and blockchain for enhanced security and accountability.

REFERENCES

[1] Thapliyal, S., Wazid, M., Singh, D. P., Chauhan, R., Mishra, A. K., & Das, A. K. (2024). Secure Artificial Intelligence of Things (AIoT)-enabled authenticated key agreement technique for smart living environment. *Computers and Electrical Engineering*, *118*, 109353.

[2] Nadifi, Z., Ouaissa, M., Ouaissa, M., Alhyan, M., & Kartit, A. (2025). Security, privacy, and trust in IoT networks. In *Artificial Intelligence for Blockchain and Cybersecurity Powered IoT Applications* (pp. 19–29). CRC Press.

[3] Adam, M., Hammoudeh, M., Alrawashdeh, R., & Alsulaimy, B. (2024). *A Survey on Security, Privacy, Trust, and Architectural Challenges in IoT Systems*. IEEE Access.

[4] Taleb, T., Samdanis, K., Mada, B., Flinck, H., Dutta, S., & Sabella, D. (2017). On multi-access edge computing: A survey of the emerging 5G network edge cloud architecture and orchestration. *IEEE Communications Surveys & Tutorials*, *19*(3), 1657–1681.

[5] Sadhu, P. K., Yanambaka, V. P., & Abdelgawad, A. (2022). Internet of things: Security and solutions survey. *Sensors*, *22*(19), 7433.

[6] Singh, A., & Chatterjee, K. (2017). Cloud security issues and challenges: A survey. *Journal of Network and Computer Applications*, *79*, 88–115.

[7] Williams, P., Dutta, I. K., Daoud, H., & Bayoumi, M. (2022). A survey on security in internet of things with a focus on the impact of emerging technologies. *Internet of Things*, *19*, 100564.

[8] Ouaissa, M., & Ouaissa, M. (2020, September). Cyber security issues for IoT based smart grid infrastructure. In *IOP Conference Series: Materials Science and Engineering* (Vol. 937, No. 1, p. 012001). IOP Publishing.

[9] Krahmann, E. (2005). Security governance and networks: New theoretical perspectives in transatlantic security. *Cambridge Review of International Affairs*, *18*(1), 15–30.

[10] Boeding, M., Boswell, K., Hempel, M., Sharif, H., Lopez Jr, J., & Perumalla, K. (2022). Survey of cybersecurity governance, threats, and countermeasures for the power grid. *Energies*, *15*(22), 8692.

[11] Jangjou, M., & Sohrabi, M. K. (2022). A comprehensive survey on security challenges in different network layers in cloud computing. *Archives of Computational Methods in Engineering*, *29*(6), 3587–3608.

[12] Mawgoud, A. A. (2020, February). A survey on ad-hoc cloud computing challenges. In *2020 international conference on innovative trends in communication and computer engineering (ITCE)* (pp. 14–19). IEEE.

[13] Anu, V. (2022). Information security governance metrics: a survey and taxonomy. *Information Security Journal: A Global Perspective*, *31*(4), 466–478.
[14] AlGhamdi, S., Win, K. T., & Vlahu-Gjorgievska, E. (2020). Information security governance challenges and critical success factors: Systematic review. *Computers & Security*, *99*, 102030.
[15] Alreemy, Z., Chang, V., Walters, R., & Wills, G. (2016). Critical success factors (CSFs) for information technology governance (ITG). *International Journal of Information Management*, *36*(6), 907–916.
[16] Mikalef, P., Boura, M., Lekakos, G., & Krogstie, J. (2020). The role of information governance in big data analytics driven innovation. *Information & Management*, *57*(7), 103361.
[17] Javadpour, A., Wang, G., & Rezaei, S. (2020). Resource management in a peer to peer cloud network for IoT. *Wireless Personal Communications*, 115(3), 2471–2488.
[18] Kumar, R., & Goyal, R. (2019). On cloud security requirements, threats, vulnerabilities and countermeasures: A survey. *Computer Science Review*, *33*, 1–48.

Chapter 7

Enhancing urban safety

AI-driven security solutions for smart cities

Harika Koormala, Kishor Kumar Reddy C, Vasavi Sravanthi Balusa, Nikitha Jillapalli, and Marlia Mohd Hanafiah

7.1 INTRODUCTION

A smart metropolis is an urban environment that makes use of generation to increase performance and enhance the exceptional of life of its residents. Monitor data and improve digital systems to improve the well-being of community members. To increase sustainability and strengthen careers in the city, smart city development is a movement driven by many sectors of urban society [1]. The six key pillars of a smart city are: social impact, intelligence policy awareness, benchmarking and best practices, smart city ecosystem, creativity, and innovation [2]. The concept of smart cities first emerged in the 1990s when attention was paid to the impact of ICT on new infrastructure within cities [3]. Smart cities use interconnected devices, sensors, and artificial intelligence (AI) to collect and analyze data, which helps to optimize in various areas of life, including traffic management energy efficiency public safety, waste management, and much more. Safety is one of the major concerns for people living in big cities and everyone wants to feel completely safe while walking around each day [4]. Smart cities leverage AI-driven technologies to enhance the safety and security of urban environments. This includes smart surveillance systems. Predictive analytics can help prevent crime, while emergency response systems and cybersecurity measures play a crucial role in protecting city infrastructure. The objective is to guarantee the security of the city, more adaptable and responsive to the challenges that emerge. At the same time, the well-being and safety of all residents is guaranteed. AI technology can enable, evaluate, and interpret large amounts of data from multiple sources to identify disease and support clinical decision-making [5]. Components of a smart city include smart agriculture, smart city services, smart homes, smart infrastructure, smart industry, smart energy, smart health, smart security, smart transportation, smart parking, and smart environment. These components are illustrated in Figure 7.1.

AI algorithms consider historical crime information to become aware of patterns and predict capacity hotspots for criminal interest. By using gadget learning, AI models can assist regulation enforcement companies to allocate

 DOI: 10.1201/9781003606307-7

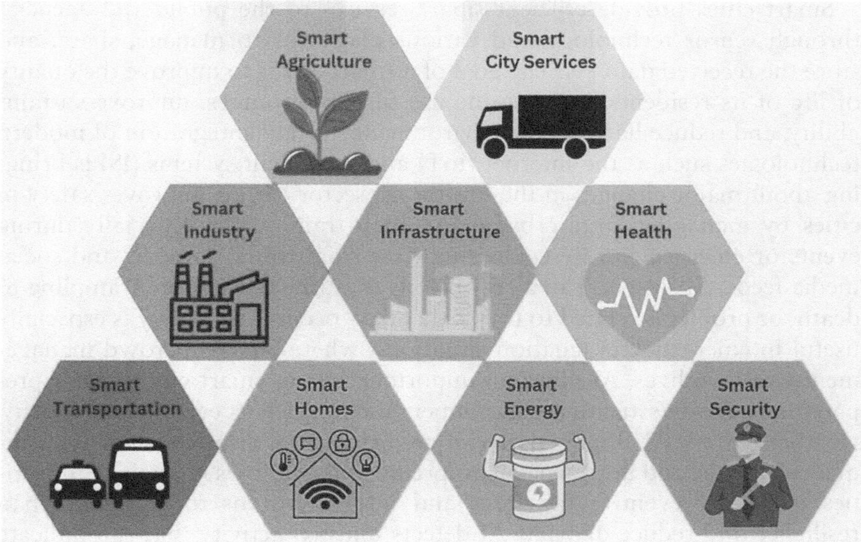

Figure 7.1 Components of a smart city.

sources extra efficiently, probably preventing crimes. Predictive policing empowers security employees to cognizance on high-hazard regions, improving proactive rather than reactive responses. AI-enabled cameras and PC imagining and vision technology allow for real-time monitoring and evaluation of public areas. These structures can come across suspicious conduct, apprehend license plates, or even discover individuals from big video feeds. This functionality enables fast detection of criminal sports or identifying threats in real-time, allowing for faster responses from law enforcement organizations. The impact of AI on our daily tasks is increasing every day [6]. AI is rapidly changing the nature of our daily tasks. It influences traditional approaches to human thoughts and interactions with the environment [6]. AI plays an important function in streamlining emergency response offerings. By examining huge quantities of statistics from sensors, social media, and verbal exchange networks, AI can pinpoint the precise vicinity and scale of incidents such as fires, injuries, or natural screw ups. This permits emergency responders to arrive on the scene quicker with appropriate assets, minimizing damage and loss. As smart towns become more interconnected, the hazard of cyber threats will also increase. AI-driven cyber security systems assist in guarding vital infrastructure by identifying and mitigating potential threats, which include data breaches, network vulnerabilities, and unauthorized access. Machine learning algorithms analyze patterns in network traffic to detect anomalies and provide predictive insights, helping to prevent attacks before they can compromise city services.

Smart cities provide efficient smart services to the public and agencies through sensor technology and various platforms to manage, share, and store the received data [7]. The goal of a smart city is to improve the quality of life of its residents, increase the use of city resources, improve sustainability, and reduce harm to the environment [8]. The integration of modern technologies such as the Internet (IoT) and intelligent systems (IS) is bringing about major changes in the healthcare sector [9]. AI improves safety in cities by managing large crowds and high traffic areas especially during events or emergencies. By evaluating data from traffic cameras and social media feeds, AI systems can alert officials to traffic congestion, trampling to death, or problems related to traffic that may occur. This ability is especially useful in emergency evacuation situations, where effective crowd management can save lives. AI plays an important role in smart city disaster preparedness. By investigating environmental data such as earthquake activity, weather pattern AI climate data can predict natural disasters such as earthquakes, floods, and hurricanes. This predictability allows municipal authorities to take preventive measures and inform citizens to increase urban resilience and reduce damage. AI detects unusual activity that can indicate danger. Smart city systems leverage machine learning. It can recognize specific movement action patterns in different parts of the city and if an anomaly occurs, such as an unexpected object in a densely populated area or abnormal movement in sensitive places, the system can then alert security personnel. This allows them to quickly assess and deal with potential threats. AI-powered systems can play a role in public health and safety. This is especially true in detecting and managing disease outbreaks or health crises. The goal of smart healthcare is to leverage technology and data to develop a more proactive, predictable, and personalized approach to healthcare management [10]. For example, during a pandemic, AI helps monitor crowd density, enforce social distancing, and track health information, which contribute to a safer urban environment. Additionally, AI can support mental health hotlines and emergency services by examining calls for potential crises and providing appropriate support. AI enables the use of drones and autonomous vehicles for surveillance and patrolling in hard-to-reach or large urban areas. Drones equipped with AI-powered cameras can cover large areas quickly, sending real-time images and data to security control centers. This approach is especially effective for keeping an eye on occurrences, boundaries, or regions that need little to no human involvement, enhancing protection and effectiveness. Residents should be trained and supported to actively participate in achieving SC city's future mission, vision, and short- and long-term plans through smart applications (such as smart open spaces that support smartphone technology).

Table 7.1 highlights the essential AI technologies used in urban security, their descriptions, and applications. Machine mastering, predictive analytics, PC vision, NLP, and AI-powered drones play important roles in the security infrastructure of smart cities. By analyzing historical and real-time data,

Table 7.1 Key AI technologies in urban security

Technology	*Description*	*Applications in urban security*
Predictive Policing	AI algorithms analyze crime data to forecast hotspots	Strategic resource allocation, crime prevention
Intelligent Surveillance	AI-enabled cameras detect and analyze suspicious activities in real-time	Public surveillance, threat detection, traffic management
Facial Recognition	Identifies individuals by matching facial features with databases	Law enforcement, event security
IoT and Sensor Networks	Collects data from traffic and environmental sensors	Real-time monitoring, hazard detection
Predictive Analytics	Forecasts patterns for proactive responses	Crime prediction, disaster preparedness
Collaborative Intelligence	Combines AI with human decision-making for flexible responses	Emergency response, cross-departmental coordination

machine learning models enhance decision-making, predict crime hotspots, and enable efficient resource allocation. Predictive analytics helps control traffic, crowd manage, and emergency response, enhancing city safety. Computer vision-powered surveillance identifies unusual actions, helps facial reputation, monitors visitors, and detects threats like deserted items. NLP analyzes social media, emergency calls, and online boards, detecting potential threats and public sentiment and allowing proactive responses. Drones geared up with AI cameras provide expansive, actual-time surveillance, reaching difficult areas, helping in crowd management, and identifying hazards such as gas leaks or fires. These autonomous patrols offer rapid reaction and actual-time statistics, enhancing security around critical infrastructure and assisting law enforcement efforts. Together, AI technology allows cities to balance safety and privacy, respond to incidents rapidly, and maintain public safety in high-density urban areas, contributing to resilient and responsive smart metropolis ecosystems.

7.2 AI APPLICATIONS IN URBAN SAFETY

AI-powered video surveillance systems are at the vanguard of current safety techniques in clever towns, providing automated evaluation and real-time tracking to locate, determine, and respond to capability threats. These systems go beyond conventional surveillance by utilizing advanced technologies such as machine learning, computer vision, and data analytics to identify patterns, detect anomalies, and enhance public safety.

Smart cities are becoming more and more of interest among governments, researchers, and industry [11]. Smart city governors can use CCTV cameras, environmental sensors, charging stations, electronic signage, Wi-Fi, and traffic management systems to collect, manage, and transfer data for decision-making [12].

AI technology is revolutionizing clever metropolis security with automated chance detection, predictive analytics, and real-time signals. Intelligent video surveillance detects suspicious activities together with loitering, unauthorized get right of entry to, or abnormal actions, triggering on the spot alerts. Facial reputation identifies humans of hobby in actual time, aiding in suspect monitoring and locating lacking men and women. AI-based crime prediction fashions examine ancient information and external elements to forecast crime hotspots, enabling optimized aid allocation. Additionally, anomaly detection pinpoints unusual actions, alerting government to capacity security threats. License plate popularity assists in site visitors control and monitoring cars linked to crook sports, even as crowd density monitoring prevents overcrowding at activities. For catastrophe management, AI systems monitor environmental hazards like smoke and flooding, offering early warnings and disaster mitigation. Predictive insights generated from historical information enhance proactive safety strategies through highlighting high-threat times and regions.

Figure 7.2 highlights the concept of facial recognition technology, showcasing a human face with a network of interconnected points, symbolizing the application of AI in recognizing, and analyzing facial features for various purposes. Facial recognition and biometric systems support secure access to restricted areas and enable seamless identity verification of public services and contactless payments. AI optimizes traffic flow and congestion management by predicting congestion patterns, detecting accidents, and providing smart parking solutions to increase pedestrian and vehicle safety.

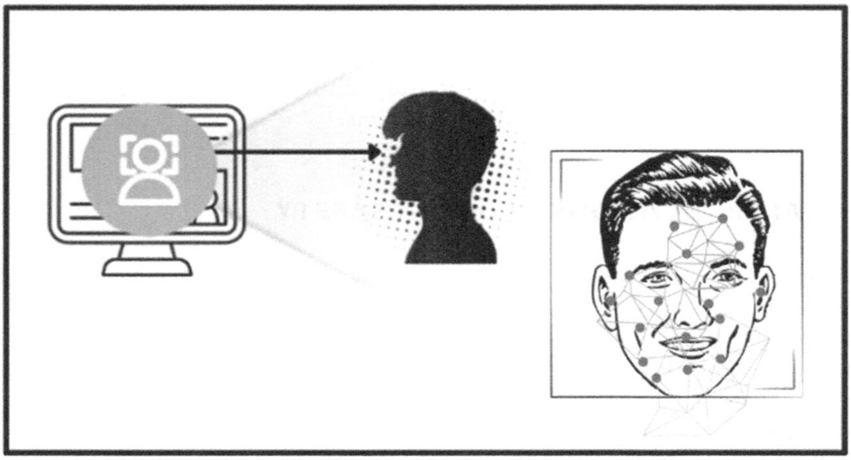

Figure 7.2 Facial recognition.

Table 7.2 Applications of computer vision in smart cities

Application	*Purpose*	*Examples*
Automated Surveillance	Detects suspicious activities without human intervention	Monitoring public spaces for security threats
Facial Recognition	Identifies individuals in public spaces	Tracking suspects, locating missing persons
License Plate Recognition (LPR)	Monitors vehicles for traffic law enforcement	Tracking stolen vehicles, congestion management
Behavior Analysis	Detects unusual movements or activities	Identifying unattended objects, suspicious behavior
Crowd Management	Monitors crowd density and movement patterns	Preventing overcrowding at events, evacuation routes
Environmental Monitoring	Detects hazards like smoke and fire	Early fire detection in high-density areas

Real-time data analysis to coordinate AI responses to emergencies recommended evacuation routes and distribute resources efficiently. The basic function of a smart city emergency management system is timely concept-based emergency processing in response to known critical situations [13]. NLP-powered social media monitoring helps officials monitor potential threats and public sentiment. Increase situational awareness with integrated data-driven insights AI-powered security systems make smart cities safer, help them respond better and enable them to deal with emergencies.

Table 7.2 reveals that computer vision enables automated surveillance and behavior analysis, as well as how to check the environment to improve public safety.

7.3 DATA AND SECURITY INFRASTRUCTURE IN SMART CITIES

Smart cities integrate Big Data and IoT to continuously collect and analyze huge amounts of data from connected devices such as sensors, cameras, smart infrastructure, etc. Due to the creation of large amounts of data, the use of smart devices therefore requires large data storage capacity. In this context, Big Data generation has replaced traditional data processing methods [14]. IoT systems refer to a growing network of digital sensors, smart appliances, and smart home appliances [15]. These data sources provide insights that help city officials increase public safety, increase efficiency in resource allocation, and improve emergency response times. By connecting devices and systems, cities can achieve a proactive, data-driven approach to urban safety. IoT devices such as smart cameras, microphones,

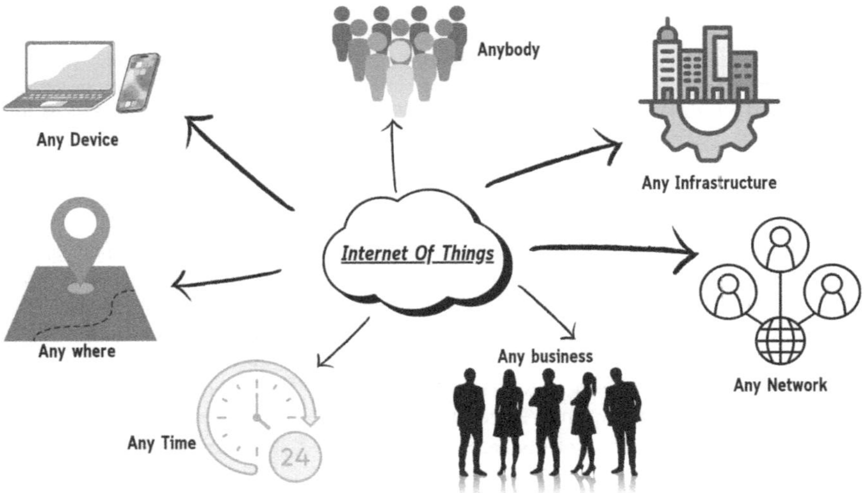

Figure 7.3 IoT concept.

and environmental sensors collect data from different locations in the city continuously. Big Data analytics processes this data to provide real-time insights, helping officials monitor urban areas for potential security threats. IoT helps create flexible and responsive production environments [16].

Figure 7.3 illustrates the concept of the Internet of Things (IoT), showing how it connects various elements like devices, locations, people, infrastructure, networks, businesses, and time. Arrows point from each element toward the central "Internet of Things," emphasizing its integration across different aspects.

Big Data and IoT devices together analyze historical and real-time data, identify patterns, and predict potential crime hotspots. Predictability allows authorities to anticipate criminal activity, which in turn enhances their efforts in preventive policing. IoT sensors embedded in roads and public areas track the movement of vehicles and pedestrians, while Big Data analytics interprets this data to prevent congestion, manage crowd density, and ensure smooth movement of IoT devices such as air quality sensors, temperature gauges, and water level monitors providing real-time information about the environment. Big Data analytics evaluates this information and identifies potential dangers such as increased pollution, fire, and flood risks. In recent years, digital water meters have been utilized to collect and transmit data on water usage. They provide real-time information about water consumption, enabling more efficient water management [17]. A combination of machine learning (ML), deep learning (DL), and data analysis (DA) concepts is used to manage the overall wastewater treatment process and support it at the convenience of the user [18].

During an emergency, IoT devices such as connected emergency alarms, public address systems, and mobile devices play a crucial role in alerting authorities and the public. It provides real-time data that can be used to coordinate responses. Big Data processes this information. It helps in making better decisions and planning resources.

Many governments around the world support smart city projects integrated with Big Data analytics to achieve sustainable urban development [19]. By integrating Big Data and IoT, cities can allocate resources based on real-time data effectively by analyzing patterns from various data sources. Authorities can identify high-risk areas or peak periods for criminal activity. Utilizing resources where they are needed most, IoT and Big Data integration also play an important role in cyber security. Sensors and IoT devices track network traffic. Meanwhile, Big Data analytics detects anomalies that may signify cyber threats, ensuring the protection of critical infrastructure. IoT devices track public health indicators such as air quality, noise levels, and radiation, providing data for Big Data systems to analyze and detect health risks. This integration is critical to ensuring a healthy and safe environment especially in densely populated urban areas. Real-time data collection and analysis continuously collect and process data from multiple sources such as IoT sensors, CCTV, social media, mobile applications, etc. This capability allows municipal authorities to monitor municipal activities, detect abnormalities, and proactively respond to security threats, thus enabling public safety in smart cities and improving operational efficiency.

Real-time data collection and analysis is a key component of smart city security, assisting urban areas in tracking and reacting to events that take place. Intelligent cities utilize sensors, cameras, and interconnected IoT devices to gather data in real time from public areas, traffic networks, and to assess the environment. AI algorithms process this data instantly, it identifies patterns, and detects anomalies that may signal potential security threats. For example, real-time data from traffic cameras can instantly adjust the timing of signals, avoid traffic build-up, and ensure entry for emergency responders. Crowd monitoring systems in high-density areas can analyze pedestrian flow to prevent overcrowding and ensure public safety. In addition, environmental sensors will alert officials to dangers such as fire or increased pollution; this makes it possible to intervene promptly. This real-time data infrastructure strengthens city safety by providing actionable insights, increase responsiveness, and enable proactive management of urban challenges.

Smart cities rely on interconnected devices, IoT sensors, and AI-driven analytics, making robust cyber security essential to protect data, infrastructure, and public safety. Network safety features, like encryption and multi-component authentication, protect IoT gadgets against unauthorized access. AI-based threat detection systems analyze network activity in real time to prevent intrusions, while data protection protocols ensure encryption and privacy compliance. End-factor security, along with everyday updates and device authentication, mitigates vulnerabilities across allotted devices. In cyber incidents, AI-driven response systems isolate breaches, assist recuperation, and minimize downtime. Critical infrastructure calls for continuous monitoring to prevent sabotage, while fact facilities and cloud offerings need strict get entry to control. Educating employees on cyber security quality practices reduces human blunders, though demanding situations continue to be in scaling these answers, defensive privateness, and adapting to evolving threats.

A general three-level architecture can generally be applied to general smart city concepts at the lowest level. Urban infrastructure refers to the physical objects that exist in a city, such as traditional elements like bus stops and traffic lights, or future objects that may occur. This could be things like drones and autonomous ground vehicles. At the highest level of three-layer architecture are smart city services. These range from electric vehicle charging services and parking services to travel applications and weather information services [20].

Figure 7.4 depicts a three-tier smart city architecture, where the "digital layer" connects urban infrastructure with smart city services. This layer processes data using sensors, communication infrastructure, and actuators to enable services like parking, waste management, and weather monitoring.

AI-powered smart city systems rely on the collection of vast amounts of data from sensors, cameras, and mobile apps, raising significant privacy and ethical concerns. Public surveillance creates a feeling of constant surveillance, which may violate personal freedom. Anonymizing data by limiting it to high-risk areas can help protect privacy. Data retention policies are important to prevent misuse of personal information. This includes clear

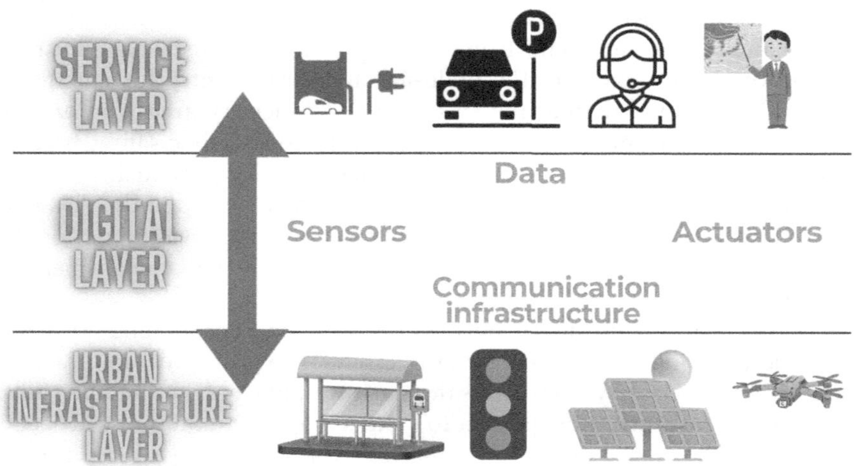

Figure 7.4 Smart city architecture.

rules about storage, access, and deletion. Strong data security measures such as encryption and auditing help reduce the risk of breach and protect sensitive data from cyberattacks.

From an ethical perspective, Transparency and consent are important. This concern arises from the possibility that the public may not fully comprehend the extent of data collection. Trust can be built by informing the public and giving them the option to leave. Bias in AI algorithms is another concern. Biased information may result in the unjust treatment of specific groups. Regular inspections and collection of information promote objectivity in AI applications. Involving humans in important decisions can ensure decision-making accountability, reduce errors, and build public trust. There is also the possibility of surveillance violations that could affect civil liberties. Independent review and rigorous guidelines are essential. Finally, cities should promote public dialogue and engagement in AI policy to promote trust and align initiatives with citizen expectations. Smart city governors can use CCTV cameras, environmental sensors, charging stations, electronic signs, Wi-Fi, and traffic management systems to collect, manage, transfer data, and use it for decision-making.

7.4 CASE STUDIES: AI-POWERED SECURITY IN GLOBAL SMART CITIES

7.4.1 AI security solutions in Singapore

Singapore has established itself as a leading smart city by deploying AI-powered safety solutions in urban areas. To increase public safety and operational efficiency, the city government uses AI-powered surveillance systems, including smart cameras with facial recognition capabilities, to inspect areas with heavy traffic such as airports, public transport stations and commercial centers. These systems allow real-time tracking of persons of interest, helps law enforcement agencies respond quickly to potential threats and crowded locations reduce distractions in the field.

In addition to tracking, Singapore is also using predictive analytics to improve resource allocation and prevent crime. By analyzing historical crime data and environmental factors, AI algorithms identify potential hotspots and recommend the appropriate deployment of law enforcement personnel. This proactive approach not only improves crime prevention, but it also helps police and emergency responders manage high-risk areas and respond quickly to incidents.

Singapore's AI-driven solutions also extend to environmental monitoring for disaster preparedness. Sensors and data analytics platforms track air quality, water levels, and weather patterns and provide early warning about natural disasters or public health risks. It reinforces Singapore's commitment to creating an improved urban environment which balances safety with efficient public services.

Table 7.3 AI-powered security in Singapore

Aspect	*Details*
Adoption of AI in Security	Singapore has implemented AI-driven security solutions across urban areas, especially in high-traffic zones such as airports, transit stations, and commercial centers.
Technology Used	AI-powered surveillance systems with facial recognition capabilities; predictive analytics to identify crime hotspots; and environmental monitoring sensors for disaster preparedness.
Objective	To enhance public safety, optimize law enforcement resource allocation, and improve disaster readiness.
Methodology	– Real-time tracking through smart cameras for quick threat response. – Predictive analytics for proactive policing. – Environmental data monitoring for early warning on hazards and public health risks.
Key Benefits	– Faster law enforcement response times. – Improved crime prevention through targeted policing. – Enhanced preparedness for natural disasters and environmental risks.
Environmental Monitoring	Sensors track air quality, water levels, and weather, enabling early warnings and preventive measures for potential disasters or health hazards.
Challenges	Balancing security with privacy concerns due to extensive use of surveillance and data collection.
Current Impact	Singapore's comprehensive AI-driven approach promotes a safe, resilient urban environment that balances advanced security with efficient public services.

Table 7.3 summarizes Singapore's AI-powered security initiatives, including technologies used, key objectives, benefits, and current impact.

7.4.2 Predictive policing in the United States

In the past few years, predictive policing techniques are being actively used to combat crime in many cities in the United States. Using AI algorithms and data analysis, law enforcement agencies analyze historical crime data to identify patterns and predict crime hotspots. In cities such as Los Angeles and Chicago, a system has been used to allocate resources efficiently. It directs police to areas where criminal activity is likely to happen. This targeted approach allows employees to employ more strategies with minimal planning, which can help prevent crimes before they escalate.

One of the most widely known predictive policing tools, PredPol, analyzes data related to time, location, and type of past crimes. This data creates a predictive map that helps police departments predict possible crimes in the near future. By focusing on specific areas during times of high-risk, law enforcement aims to increase visibility and protection, consequently helping to make the community safer. However, predictive policing raises ethical concerns about bias and fairness. Critics argue that algorithms trained on past crime data can reinforce existing biases, which disproportionately affects some communities. To address these concerns, some cities in the United States are improving transparency, conducting regular bias checks, and creating guidelines to ensure that predictive healthcare models are applied equitably. Despite these challenges, predictive healthcare remains an impressive tool in the United States. The aim is to strike a balance between proactive crime prevention and responsible use of AI.

Table 7.4 provides an overview of key points regarding predictive policing in the United States, including technology, objectives, benefits, ethical concerns, and mitigating actions.

Table 7.4 Predictive policing in the United States

Aspect	*Details*
Adoption in U.S. Cities	Several U.S. cities, including Los Angeles and Chicago, have implemented predictive policing techniques to address crime proactively.
Technology Used	AI algorithms analyze historical crime data to predict potential crime hotspots. Tools like PredPol use data on time, location, and type of past crimes to create predictive maps.
Objective	To enable more strategic police resource allocation by identifying high-risk areas and times, aiming to prevent crimes before they occur.
Methodology	Predictive models analyze crime patterns and environmental factors, guiding police presence to areas with a higher probability of criminal activity.
Key Benefits	– Efficient resource allocation. – Increased police visibility in high-risk areas. – Potential deterrence and reduction of crime rates.
Ethical Concerns	– Risk of reinforcing biases from historical crime data, potentially leading to unfair targeting of specific communities. – Concerns about transparency and fairness in AI-driven decisions.
Mitigating Measures	Some cities are introducing transparency measures, conducting regular bias audits, and establishing guidelines for equitable application of predictive policing models.
Current Impact	Predictive policing remains influential in the United States, focusing on crime prevention while attempting to balance effectiveness with responsible AI use.

7.5 CHALLENGES AND LIMITATIONS OF AI IN SMART CITY SECURITY

1. Data quality and computational needs: AI systems in urban security require high-quality data, but there are inconsistencies from sources like sensors and social media. This often affects the accuracy and reliability of the model.
2. Integration challenges: Integrating AI into existing urban infrastructure is complex. This is especially true in cities with legacy systems. This creates barriers to smooth real-time data processing.
3. Scalability concerns: Scaling AI systems to support increasing data volumes and expanding urban infrastructure requires large amounts of computing power and secure data storage.
4. Cyber security risks: AI systems themselves are vulnerable to hacking and data theft. This requires flexible and adaptive cyber security measures.
5. Balancing Security and Privacy: The most important obstacle "Information sharing is not enough," privacy protection strategies such as de-identification needs to be addressed to achieve the smart city concept [19].

Figure 7.5 shows various challenges faced by smart city IOT systems like network access, robustness, data heterogeneity, security, mobility, privacy, data leakage, scalability, etc.

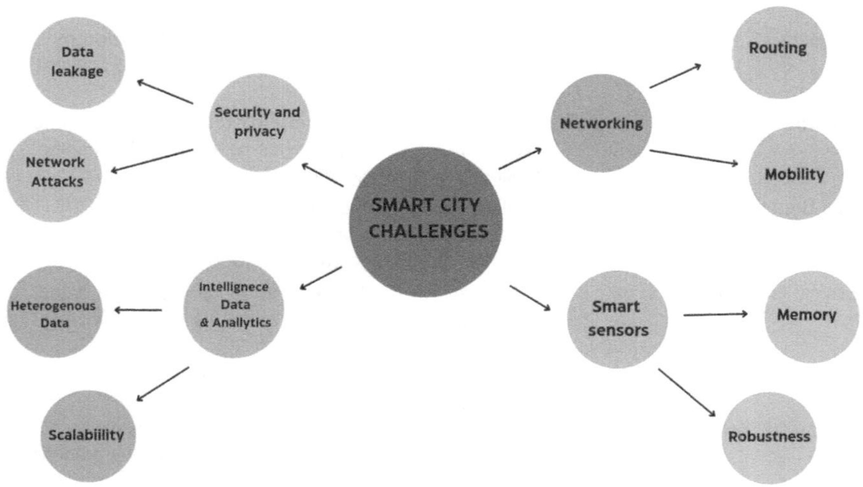

Figure 7.5 Smart city challenges.

6. The main challenge at the level of device identification, in a smart city IoT edge computing environment, is the selection of reliable participants. This is because some IoT smart devices may not be reliable. Some smart IoT devices can cause harmful damage to networks or services and affect the service quality of the system [21].
7. Ethical Dilemmas: AI can create bias in surveillance, unfairly targeting specific groups and lead to violations by officials. This raises ethical concerns about objectivity and independence.
8. AI Surveillance Responsibilities: To limit abuse, clear guidelines, regular audits, and ethical oversight are essential to ensure that AI complies with democratic principles and human rights.
9. Regulatory gaps: Existing privacy and data laws often do not address AI's unique challenges, creating legal uncertainty that may slow AI adoption in smart cities.
10. Inconsistent regulations across jurisdictions: Varying privacy and AI regulations around the world make standardization difficult, complicating cross-border AI solutions and its use in maintaining security in the city.

Table 7.5 outlines the major technical and ethical challenges in implementing AI-driven security in smart cities, including data quality and privacy concerns.

Table 7.5 Challenges and limitations of AI in smart city security

Challenge	*Description*	*Impact*
Data Quality	Inconsistent data sources affect model accuracy and reliability	Reduced model effectiveness
Integration Complexity	Difficult to merge AI with legacy infrastructure	Delays in real-time data processing
Scalability	Large data volumes and expanding infrastructure demand high computational power	Higher costs and resource demands
Cyber Security Risks	AI systems are vulnerable to hacking and data breaches	Increased need for resilient cybersecurity measures
Balancing Security and Privacy	Privacy concerns due to extensive surveillance require transparent policies	Public mistrust
Ethical Dilemmas	Potential biases in AI algorithms may lead to unfair treatment	Risks of social inequality and bias

7.6 FUTURE TRENDS IN AI FOR SMART CITY SECURITY

The advent of 5G and edge computing is revolutionizing AI applications in smart cities. It facilitates faster data processing and reduces latency. Edge computing plays an intermediary role between IoT devices and cloud computing environments to speed up data analysis [22]. High-speed 5G connectivity enables AI-based systems like real-time video surveillance, self-driving vehicles, and predictive maintenance to run smoothly, transmitting and processing data in milliseconds. Edge computing moves data processing closer to the source (e.g., IoT devices and cameras), reducing reliance on centralized data centers and enabling faster, localized decision-making. Data centers helps make local decisions faster; this is especially important in critical applications that require real-time response, such as traffic management or emergency response. Together, 5G and edge computing are increasing the capability. Scale the reliability of AI systems, support more IoT devices, and make cities more resilient. Respond quickly to urban challenges and enhance resource management.

As quantum computing technology advances, AI-powered cybersecurity has become critical in protecting smart city infrastructure and sensitive data. Quantum computers pose risks to traditional encryption methods. This may make it easier for cybercriminals to breach data security. AI can counter these threats with dynamic, adaptive cyber security. It quickly detects abnormalities and responds to violations. Machine learning models trained to identify quantum threats can analyze network traffic patterns in real time. As AI-powered predictive analytics can predict vulnerabilities and recommend proactive protection, AI will play a key role in post-quantum encryption. Develop cryptographic algorithms that are resistant to quantum-based attacks. This combination of AI and cyber security will help protect critical infrastructure such as the power grid and transportation networks. This will ensure that smart city systems have an architecture that remains resistant to the complex threats posed by quantum computing. It is a key component of the strategy for a sustainable energy future. This is because it can not only facilitate the integration of renewable energy sources and the electrification of transportation, but also enable value-added services related to new energy [23].

Another transformative trend is the upward push of collaborative intelligence and independent, self-sustaining protection systems in clever cities. Collaborative intelligence permits a synergy among human decision-makers and AI systems, ensuring that city safety measures benefit from AI's speed and information insights while preserving human judgment. Collaborative robots, additionally called cobots, have arisen as a revolutionary technological advancement aimed at operating along with human operators, consequently augmenting production efficiency and potential [24]. In business operations, for example, AI can optimize routes and prognosticate traffic, while mortal drivers make environment-sensitive adaptations. Autonomous protection systems, then again, leverage advancements in AI, IoT, and robotics to detect and reply to threats independently. AI-powered drones and predictive algorithms permit these structures to patrol regions, expect dangers, and adapt their responses through the years, all even as being powered sustainably by means of sun electricity and aspect computing. Together, collaborative intelligence and autonomous systems are shaping a destiny of resilient, self-sufficient city protection, where AI enhances safety while remaining flexible to the dynamic needs of city environments.

7.7 CONCLUSION

Smart cities leverage emerging technologies such as Big Data, Internet of Things (IoT), Cloud Computing, and artificial intelligence (AI) to improve public service management [19]. Using AI in smart city security systems changes city safety, allowing for faster and more proactive responses to threats and improving overall resource management. AI predictive policing helps cities in crime prevention, traffic management, and improving emergency responses through applications such as smart video surveillance and real-time data analysis. Both operational effectiveness and public safety could be greatly enhanced by these developments, creating a smarter and more flexible urban environment.

However, this change comes with significant challenges that must be addressed. Technical barriers such as data quality, scalability, and cybersecurity risks require a strong infrastructure and continuous optimization. Additionally, it is essential to balance security with privacy and ethical standards. This is to avoid AI oversight, bias, and abuse, which can lead to public distrust. Regulatory gaps further complicate the landscape. This is because existing frameworks often lack in meeting the unique needs of AI-driven technologies.

To completely harness AI's capacity in city safety, cities need a balanced technique that includes sturdy cyber security measures, obvious information practices, ethical oversight, and adaptable regulations. Collaboration among policymakers, era vendors, and groups can be crucial in building agreement with and fostering responsible AI use. By aligning innovation with privateness, fairness, and accountability, smart towns can leverage AI to create secure, inclusive, and sustainable environments that serve the well-being of all residents.

REFERENCES

[1] Yang J., Kwon Y. and Kim D., "Regional Smart City Development Focus: The South Korean National Strategic Smart City Program," *IEEE Access*, vol. 9, pp. 7193–7210, 2021, doi: 10.1109/ACCESS.2020.3047139

[2] Lytras M. D., Visvizi A., Torres-Ruiz M., Damiani E. and Jin P., "IEEE Access Special Section Editorial: Urban Computing and Well-Being in Smart Cities: Services, Applications, Policymaking Considerations," *IEEE Access*, vol. 8, pp. 72340–72346, 2020, doi: 10.1109/ACCESS.2020.2988125

[3] Singh, A. Solanki, S. K. Sharma, A. Nayyar and Paul A., "A Decade Review on Smart Cities: Paradigms, Challenges and Opportunities," *IEEE Access*, vol. 10, pp. 68319–68364, 2022, doi: 10.1109/ACCESS.2022.3184710

[4] Salama, R., Al-Turjman, F., Culmone, R.," AI-Powered Drone to Address Smart City Security Issues." In: Barolli, L. (eds) *Advanced Information Networking and Applications. AINA* 2023. vol 655. Springer, Cham. doi: 10.1007/978-3-031-28694-0_27

[5] Rangarajan Deepti, Rangarajan Aarti, C. Kishor Kumar Reddy and Srinath Doss, "Exploring the Next-Gen Transformations in Healthcare Through the Impact of AI and IoT." In "*Intelligent Systems and IoT Applications in Clinical Health*" (pp. 73–98). IGI Global. doi: 10.4018/979-8-3693-8990-4.ch004

[6] Ullah, Zaib, Al-Turjman Fadi, Mostarda Leonardo and Gagliardi Roberto. "Applications of artificial intelligence and machine learning in smart cities." *Computer Communications*, 154 (2020): 313–323. doi: 10.1016/j.comcom.2020.02.069

[7] Talebkhah M., Sali A., Gordan M., Hashim S. J. and Rokhani F. Z., "Comprehensive Review on Development of Smart Cities Using Industry 4.0 Technologies," *IEEE Access*, vol. 11, pp. 91981–92030, 2023, doi: 10.1109/ACCESS.2023.3302262

[8] Mohamed N., Al-Jaroodi J., Jawhar I. and N. Kesserwan, "Data-Driven Security for Smart City Systems: Carving a Trail," *IEEE Access*, vol. 8, pp. 147211–147230, 2020, doi: 10.1109/ACCESS.2020.3015510

[9] Sree, M. Swathisree and C. Kishor Kumar Reddy. "Applications of Intelligent Systems and the Internet of Things in Clinical Health." In *Intelligent Systems and IoT Applications in Clinical Health*, IGI Global, doi: 10.4018/979-8-3693-8990-4.ch003

[10] Sree, Swathi, Kishor Kumar Reddy and Srinath Doss. "Smart Healthcare Innovations Using Intelligent Systems in Industry 4.0." In "Integration of AI, Quantum Computing, and Semiconductor Technology", IGI Global, 2025. doi: 10.4018/979-8-3693-7076-6.ch018

[11] Arora A., Jain A., Yadav D., Hassija V., Chamola V. and Sikdar B., "Next Generation of Multi-Agent Driven Smart City Applications and Research Paradigms," *IEEE Open Journal of the Communications Society*, vol. 4, pp. 2104–2121, 2023, doi: 10.1109/OJCOMS.2023.3310528

[12] Bokhari S. A. A. and Myeong S., "The Impact of AI Applications on Smart Decision-Making in Smart Cities as Mediated by the Internet of Things and Smart Governance," *IEEE Access*, vol. 11, pp. 120827–120844, 2023, doi: 10.1109/ACCESS.2023.3327174

[13] Costa D. G. et al., "A Survey of Emergencies Management Systems in Smart Cities," *IEEE Access*, vol. 10, pp. 61843–61872, 2022, doi: 10.1109/ACCESS.2022.3180033

[14] Talebkhah M., Sali A., Marjani M., Gordan M., Hashim S. J. and Rokhani F. Z., "IoT and Big Data Applications in Smart Cities: Recent Advances, Challenges, and Critical Issues," *IEEE Access*, vol. 9, pp. 55465–55484, 2021, doi: 10.1109/ACCESS.2021.3070905

[15] Kirimtat A., Krejcar O., Kertesz A. and Tasgetiren M. F., "Future Trends and Current State of Smart City Concepts: A Survey," *IEEE Access*, vol. 8, pp. 86448–86467, 2020, doi: 10.1109/ACCESS.2020.2992441

[16] Reddy C.K.K., Anisha P.R., Hanafah M.M., Doss S., and Lipert K.J., "*Intelligent Systems and Industrial Internet of Things for Sustainable Development*", CRC Press, Taylor & Francis, New York.

[17] Bawankar N., Kriti A., Chouhan S. S. and Chaudhari S., "IoT-Enabled Water Monitoring in Smart Cities with Retrofit and Solar-Based Energy Harvesting," *IEEE Access*, vol. 12, pp. 58222–58238, 2024, doi: 10.1109/ACCESS.2024.3392852

[18] Reddy K.K., Reddy P.Y., Hanafiah M.M. and Doss S., 'Intelligent Systems and Robotics for Wastewater Management Across India: A Study and Analysis" In "*Smart Sensors for Industry 4.0: Fundamentals, Fabrication and IIoT*." Wiley, doi: 10.1002/9781394214723.ch8

[19] Khan M. A., Siddiqui M. S., Rahmani M. K. I. and Husain S., "Investigation of Big Data Analytics for Sustainable Smart City Development: An Emerging Country," *IEEE Access*, vol. 10, pp. 16028–16036, 2022, doi: 10.1109/ACCESS.2021.3115987

[20] Sivrikaya F., Ben-Sassi N., Dang X. -T., Görür O. C. and Kuster C., "Internet of Smart City Objects: A Distributed Framework for Service Discovery and Composition," *IEEE Access*, vol. 7, pp. 14434–14454, 2019, doi: 10.1109/ACCESS.2019.2893340

[21] Wang B., Li M., Jin X. and Guo C., "A Reliable IoT Edge Computing Trust Management Mechanism for Smart Cities," *IEEE Access*, vol. 8, pp. 46373–46399, 2020, doi: 10.1109/ACCESS.2020.2979022

[22] Gheisari M., Pham Q.-V., Alazab M., Zhang X., Fernández-Campusano C. and Srivastava G., "ECA: An Edge Computing Architecture for Privacy-Preserving in IoT-Based Smart City," *IEEE Access*, vol. 7, pp. 155779–155786, 2019, doi: 10.1109/ACCESS.2019.2937177

[23] Masera M., Bompard E. F., Profumo F. and Hadjsaid N., "Smart (Electricity) Grids for Smart Cities: Assessing Roles and Societal Impacts," *Proceedings of the IEEE*, vol. 106, no. 4, pp. 613–625, April 2018, doi: 10.1109/JPROC.2018.2812212

[24] Anisha P.R., Reddy C.K.K., Nguyen N.G., Bhushan M., Kumar A. and Hanafiah M.M., "*Intelligent Systems and Machine Learning for Industry: Advancements, Challenges, and Practices.*", CRC Press.

Chapter 8

AIoT security in healthcare systems

Challenges, opportunities, and future directions

Hatem Mosa and Qasem Abu Al-Haija

8.1 INTRODUCTION

The healthcare sector has profoundly changed by combining two compelling technologies: the Internet of Things (IoT) and artificial intelligence. The integration of these two technologies, referred to as the Artificial Intelligence of Things (AIoT), has transformed healthcare systems. IoT devices connected to the Internet play a crucial role in gathering real-time data from patients while working in conjunction with artificial intelligence algorithms. This collaboration allows for comprehensive data analysis and the development of actionable plans, enabling faster and more accurate decision-making. AIoT is expected to effectively improve patients' health conditions. However, despite its many benefits and advantages, there are considerable concerns about using AIoT in healthcare. Many complex security obstacles threaten the safety of healthcare systems and patients. This chapter presents the impact of AIoT development on healthcare and the security concerns that result from its use [1].

8.1.1 The evolution of AIoT in healthcare

The incorporation of IoT technology in healthcare systems began with the proliferation of IoT and artificial intelligence in healthcare systems. Initially, the application of IoT focused on acquiring primary data using sensors. Among the first tools of the IoT are devices that track patient health metrics, such as glucometers, blood pressure meters, and heart rate monitors. The initial implementation of IoT devices effectively collected information; however, it exhibited limitations in analyzing or interpreting the acquired data. Therefore, artificial intelligence technologies have been introduced to process these vast amounts of data, which would, in turn, enable healthcare systems to generate insights and make informed decisions [2]. Over time, more advanced and complex artificial intelligence algorithms are consistently integrated into IoT networks, making them more intelligent and sophisticated.

 DOI: 10.1201/9781003606307-8

In healthcare, remote patient monitoring (RPM) was one of the earliest cases of utilizing the IoT. Using RPM makes it easier for healthcare providers to follow up on patients outside the clinical setting. Wearable sensors and home monitoring tools continuously produce data that can detect diseases and predict health risks before they become severe. Healthcare practitioners now require sophisticated systems to efficiently comprehend and evaluate the increased volume and complexity of data. In this case, artificial intelligence successfully bridges this gap by accessing data produced by healthcare systems in real time, forecasting how these events would turn out and automatically producing decisions and classifications.

AI assists and actively participates in healthcare diagnostics and treatment planning. For example, AI algorithms are used to assess magnetic resonance imaging (MRI) or computed tomography (CT) images to identify cases that contain early signs of diseases, such as cancer or neurological disorders. Some studies have shown that AI can detect the features and patterns of images that radiologists in some areas of practice cannot, allowing for more accurate image diagnosis, treatment, and even better results. This demonstrates that although IoT devices function nonstop and monitor many essential signs that improve health services delivered within a healthcare facility [3, 4], AI simultaneously continues to evolve, gaining a more critical role in healthcare, such as in drug development and tailor-made treatment. Additionally, AIoT can monitor patients' vital signs to help maintain automatic tracking and reduce the time required to process and transfer data through modern connected networks.

8.1.2 The rapid growth of AIoT in healthcare

The integration of AIoT in healthcare is notable because of the high demand for more effective and elaborate healthcare services. AI in healthcare market was estimated at $27.6 billion in 2021, and by 2030, it is anticipated to grow at an annual rate of 37.3%, according to a report by Grand View Research [5]. However, this growth can be aided by the increased need for AI-powered applications to plan diagnosis and treatment, track patients, and prevent diseases. Such care can be delivered through AIoT technologies, which allow service providers to collect and interpret data from live networks of wearable medical devices, imaging systems, and patient medical history files.

In addition, the healthcare IoT market is on an upward trend, as reported by Future Market Research [6], according to which, healthcare IoT market growth innovation will reach $534.3 billion in 2025. This growth has resulted from the increased use of connected medical devices that allow providers to track patients. Examples of this include blood pressure monitors, glucose sensors, and fitness tracker watches that can all provide crucial health monitoring.

The COVID-19 pandemic has also facilitated accelerated implementation of AIoT technologies. With the requirement for physical separation and isolation, many healthcare systems embraced telehealth and remote patient monitoring systems to cater to patients who could not visit hospitals or clinics. McKinsey & Company [7] sought to prove these estimates by reporting a 58% increase in the utilization of telemedicine in the United States during the pandemic, emphasizing the importance of AIoT tools in healthcare service delivery. AI-enhanced diagnostic applications, virtual consultations, and AI-integrated wearable technology have been critical in enabling people to receive medical attention, regardless of the issues and challenges posed by the pandemic.

Since then, significant improvements in AIoT have emerged. For example, AI-based diagnostic aids have proven helpful in medical imaging. In the research published in *The Lancet* [8], AI systems outperformed human radiologists in detecting cancer much earlier, which allowed the right treatments to be applied at the right time. Moreover, wearable devices integrated with AIoT technology are now used to monitor chronic diseases, such as diabetes, hypertension, and heart diseases.

8.1.3 Security risks and challenges in AIoT healthcare systems

Despite the enormous benefits of using AIoT in healthcare, it has also introduced technological risks to different applications in healthcare systems. The risks associated with AIoT in healthcare have been increasing owing to the increased usage of AI-powered devices. Cyber threats, data leaks, and system vulnerabilities have become severe concerns in AIoT applications in the healthcare domain.

Today, many medical devices, such as ventilators, insulin delivery systems, and pacemakers, are associated with clinical networks and have internet access. Complications due to internet-connected devices pose a serious challenge to medical infrastructure, notably when they have features such as automation and remote control. In 2017, the U.S. Food and Drug Administration (FDA) recalled nearly 50,000 infusion pumps after the exposure of vulnerabilities that allow attackers to remotely change the dosage of medication given to patients, which could have life-threatening consequences for patients who rely on these devices for vital treatments. [9]. Another case occurred within the same year, when a significant cybersecurity vulnerability attack occurred in the healthcare sector, affecting over 200,000 computers in more than 150 countries, including healthcare institutions worldwide [10]. The National Health Service in the United Kingdom suffered critically due to this hack since medical institutions could not access patients' records, which led to postponement of the treatment dose and cancellation of surgical appointments. This incident helped highlight the weaknesses of the systems in healthcare institutions. Such weaknesses require measures to safeguard the cyber domain in healthcare, protect patients' data and privacy, and ensure the necessary functioning of essential medical devices [6]. Although the aforementioned incident is not directly related to AIoT, it is worth mentioning as an example to simulate the dangers and effects of technical vulnerabilities in the healthcare sector.

One of the critical risks and challenges of using AI is that AI algorithms are not entirely transparent, as they are described as black boxes, which makes integrating such technology with the IoT in a sensitive domain such as healthcare more critical. In addition to the inability to explain the reasons behind AI algorithm decisions and final outputs, the absence of transparency also results in failure to identify compromised artificial intelligence systems. For example, these compromised systems affect patients by providing incorrect diagnoses and inappropriate treatment recommendations. In this case, the health industry is particularly affected, and these violations endanger lives [11].

8.1.4 The impact of security incidents in healthcare

The consequences of security breaches in IoT healthcare systems can be seen as catastrophic. Healthcare organizations may suffer from data breaches that affect them financially, damage their reputation, and cause a loss of credibility and patient trust. According to IBM's report on the cost of healthcare data breaches, the average is approximately $9.23 million, almost the highest of all breaches in other categories [12]. This financial aggravation results from the legal consequences imposed by the state owing to the failure to protect and preserve patient data. Moreover, security breaches in healthcare directly endanger the lives of patients. When an attacker gains access to a medical device, they may alter the treatment protocols, leading to incorrect drug doses or failure to monitor the patient through the devices correctly. For example, when an insulin pump is hacked, the treatment protocol can be changed, and the change in the doses administered to the patient may cause severe changes in the blood sugar levels, exposing the patient to danger [13].

Although AIoT positively impacts healthcare by making it more efficient and responsive to patients' needs, it has highlighted the severe security challenges that must be addressed to achieve the desired benefits of these technologies. Applying solid security standards to these technologies is necessary to address the shortcomings of medical devices connected to patients. To protect patients' data, ensure their safety, and ensure the safe use of artificial intelligence in monitoring and treating disease cases, healthcare institutions must implement security strategies for their systems and provide transparency in using artificial intelligence models while considering all patients' privacy. The future of healthcare depends mainly on achieving the right balance between creativity and maintaining security, and the need to ensure the ability and efficiency of the IoT to improve patient care without compromising their safety or privacy.

8.2 AIoT IN HEALTHCARE SYSTEMS

Initially, the goal of the IoT was to collect and share healthcare datasets for research using IoT devices and to allow those devices to communicate with each other through different types of networks. IoT has subsequently enabled the development of patient-connected devices such as wearable sensors, remote monitoring tools, and medical devices that facilitate the collection and transfer of patient data in real time. Artificial intelligence was added to these devices, making them capable of analyzing vast amounts of data, predicting the condition of patients, and predicting the risks expected to occur, which greatly helped many specialists make timely health decisions.

The combination of the Internet of Things and artificial intelligence has not only led to the development of technological devices but also the provision of care for patients and the ability to monitor them remotely by specialists. AIoT has become a solution to overcome traditional healthcare issues and challenges, such as increased waiting times, a shortage of specialists, and expensive fees, by providing better and more efficient healthcare services [7]. In this section, we examine how AIoT integrates sensors, cloud computing, AI algorithms, and communication protocols to improve healthcare services and specific applications of AIoT in healthcare systems [5].

8.2.1 Components of AIoT in healthcare systems

The components of AIoT in healthcare systems continue to develop and change periodically. These components work together, enabling each other to create comprehensive advanced healthcare applications that include the following:

- **Sensors**: Sensors are essential for collecting patient data and can detect blood pressure, temperature, blood sugar, oxygen saturation, heart rate, and other medical parameters [6]. For example, some devices are equipped with sensors that continuously monitor a patient's heartbeat. If any abnormalities or irregularities in the heartbeat are detected medical staff can be alerted take the necessary measures. By continuously transmitting data through these devices, patients suffering from chronic diseases can be monitored remotely instead of frequent physical visits to the hospital [5].
- **Cloud computing**: Cloud computing is a central platform for receiving, storing, processing, and managing data generated by remote sensors in the cloud. Cloud computing provides fast access to, retrieval of, and secure data sharing, facilitating decision-making by healthcare providers [7]. Cloud computing also transfers data collected by IoT devices to the cloud for processing and analysis, enabling healthcare providers to make decisions based on these analyses and doctors to access patient information quickly and remotely. One of the essential

features of cloud computing is its ability to expand data storage and allocated resources, allowing healthcare institutions to store and analyze vast amounts of big data such as patient information, medical tests and results, and patient images.

- **Artificial intelligence algorithms**: Artificial Intelligence algorithms can analyze data collected by IoT devices by applying machine learning and deep learning. After analyzing the data, the trained models can diagnose the disease, predict risks that may affect the patient, and determine ways to prevent them by examining current and historical patient data [14].
- **Communication protocols**: Communication protocols define how data are transmitted or exchanged, especially over a network. It permits data to be efficiently shared between IoT devices, cloud computers, and healthcare providers through Bluetooth, Wi-Fi, Zigbee, and 5G. Here, the emergence of fifth-generation technology reduces data transfer time and provides high-speed communications. By exchanging information faster and with higher accuracy, it is easier to implement artificial intelligence applications in healthcare [15].

8.2.2 Applications of AIoT in healthcare

AIoT applications in healthcare are somewhat limited, as new and innovative areas that can be utilized to benefit from this technology are constantly developing. Currently, some of the prominent areas where AIoT integration is anticipated to create significant changes in the healthcare domain are as follows:

- **Remote monitoring**: Remote monitoring is the technology used to monitor smart devices in the Internet of Things (IoT). One of the most transformative applications of AIoT is to monitor and manage chronic diseases, such as diabetes, hypertension, and cardiovascular diseases. Remote monitoring devices continuously collect patient data and transmit them to healthcare providers to make necessary decisions. For example, wearable devices monitor blood glucose levels in diabetics, and artificial intelligence algorithms analyze these data and send notifications when any abnormality is detected in any analysis ratio, which enables doctors to intervene quickly and take the necessary action [16].
- **Diagnosis**: In radiology, for example, artificial intelligence algorithms examine and analyze medical X-rays to detect pathological conditions, such as fractures, lung disease, and cancer, with a high accuracy equivalent to that of a human radiologist. For example, results showed that AI algorithms could detect early-stage lung cancer from CT scans by 94%, and this early detection improves treatment outcomes [17]. Also,

in pathology, AI helps pathologists by training them on thousands of tissue samples to use them to identify cancerous cells. Therefore, the integration between AIoT and diagnostic systems provides the ability to customize treatment based on accurate analysis processes and provides treatments that target a specific disease, which reduces the risks and side effects that may be exposed [6].

- **Surgical assistance:** Robotic systems embedded with IoT devices and AI-powered devices can aid in many surgical procedures. The Da Vinci Surgical System is an AI-based robot that uses surgical procedures to monitor patients and to make real-time decisions. Such robots can perform complex tasks in surgical operations, such as suturing and processing tissues without human intervention, thereby improving accuracy and reducing the occurrence of risks and complications [18]. In addition, robotic systems that operate with artificial intelligence are used to analyze data from IoT sensors in real time. Based on these analyses, doctors perform surgeries with higher levels of accuracy. These robots also help to adjust surgical techniques based on the patient's medical condition, thus developing the capabilities of surgical operations.
- **Elderly care:** With an increasing proportion of the elderly in populations worldwide, the demand for providing nursing homes for older people has been on the rise. AIoT played a vital role in this field by providing innovative environments that enhance the quality of life of older people, such as Smart Environments and Fall Detection Systems. Fall detection systems use IoT sensors available inside the home or on wearable devices to monitor the movement of older people. If an older adult falls, the system alerts healthcare providers or contacts emergency services to take necessary measures to ensure a rapid response [19]. In smart homes with AIoT technologies, environmental factors, such as temperature, air, and humidity, are monitored and adjusted to provide older people with a safe and comfortable environment. This integration of AIoT improves the health of older people and ensures their safety and independence [20].

AIoT technology has revolutionized healthcare to enhance patient care through many solutions such as remote patient monitoring, improved real-time diagnostic efficiency, predictive diagnostics, surgical interventions, and elderly care. By combining sensors, cloud computing, AI algorithms, and communication protocols with this tremendous advancement in technology, healthcare providers must ensure that these systems are secure and comply with legal, health, and regulatory standards while focusing on protecting patient privacy and maximizing their potential to transform healthcare delivery globally.

8.3 SECURITY CHALLENGES IN AIoT HEALTHCARE SYSTEMS

In this section, we discuss some of the security challenges that have emerged owing to the integration of IoT and AI in healthcare. These challenges are presented from more than one perspective. The challenges will be presented through detailed threat categories, technical challenges, ethical and legal considerations, and risks to which healthcare systems based on artificial intelligence and the IoT may be exposed.

8.3.1 Detailed threat categories

8.3.1.1 Ransomware attacks

Ransomware is a malicious program that prevents users and organizations from accessing files and data on their devices until a ransom is paid; thus, it poses a significant threat to infected systems. WannaCry [21] is a well-known example of ransomware. In 2017, attackers penetrated national health institutions in the United Kingdom, and this attack affected 80 institutions affiliated with the Health Services Authority, which forced most hospitals to cancel numerous appointments and surgeries, seriously impacting patient healthcare. Such incidents highlight the criticality of the security vulnerabilities affecting IoT devices in the healthcare sector. In 2022, IBM Security conducted a study showing that the healthcare sector is most targeted by cyberattacks, with an average breach of $10.1 million per incident. [12].

8.3.1.2 Data breaches

IoT systems process critical patient data, making them desirable targets for cyberattacks. In 2023, the HCA Healthcare data breach exposed the information of 11 million patients, resulting in compromised centralized systems. In AIoT systems, data breaches often exploit weak encryption protocols, insecure access controls, and insecure cloud storage [22].

8.3.1.3 Insider threats

Insider threats come from healthcare workers who abuse or misuse their access to sensitive data. Such abuse can be either intentional or accidental [23]. In 2020, San Diego hospital employees stole patient data for personal gain. Therefore, insider threats are always a concern in IoT devices, as they connect many devices and systems, making it easy to access data on one of these devices [24].

8.3.2 Technical challenges

8.3.2.1 Vulnerabilities in IoT firmware

IoT devices may have outdated software or unpatched security vulnerabilities like any device. In 2021, a group of researchers found more than 100 vulnerabilities in commonly used medical devices, such as infusion pumps and imaging systems. These devices are exposed to cyberattacks that may result in unauthorized access to devices and systems, data leakage, or alteration of device configurations and functions [25].

8.3.2.2 Communication protocols

IoT devices are based on several protocols, including the MOTT protocol, which sequences message data, and ZigBee protocol, which transfers data. Although these protocols enable effective communication, they are often unencrypted, exposing data to violations and modifications [26]. Limited resources in IoT devices are one of the main challenges that make heavy computational encryption algorithms unsuitable.

8.3.2.3 AI-training data vulnerabilities

AI algorithms in healthcare work with large amounts of big data during the model training phase, making them vulnerable to hostile attacks. Cyberattacks may target these data, cause bias, and produce false outcomes in the model. Such weaknesses threaten the reliability of AI systems in the critical healthcare field [27].

8.3.2.4 Emerging threats

Adversarial AI: Attackers feed false input to an AI, resulting in misguided and inappropriate outputs [28].

Backdoor attacks: AI systems can be injected with malicious code that can be activated once conditions are met, exposing the entire model to threat [29].

Supply chain vulnerabilities: The dependence on IoT infrastructure has many drawbacks, as vulnerabilities within the supply chain may entail hacking the healthcare ecosystem [29].

8.3.3 Ethical and legal challenges

8.3.3.1 Regulatory compliance

Compliance with regulations and laws in AIoT healthcare systems is essential as it is considered a model for compliance with the General Data Protection

Regulation (GDPR) and Health Insurance Portability and Accountability Act (HIPAA) laws. The GDPR sets strict standards to protect data from breaches, whereas HIPAA focuses on protecting patient information. Failure to comply with these laws may result in significant financial and reputational damage to healthcare providers. For example, according to HIPAA regulations, a U.S. hospital was fined $6.5 million in 2021 due to its inability to protect patient data [30].

8.3.3.2 Ethical concerns

Decision-making using AI models in healthcare raises many ethical concerns, such as:

Bias in AI models: AI is invariably biased when trained on a sample that is either unbalanced to cases or too inclusive of cases, leading to incorrect treatment outcomes [31].

Accountability: Concerns about accountability arise when considering the persons responsible for cases of errors resulting from AI decision-making, such as misdiagnosis.

Privacy: The continued use of IoT devices to monitor patients' health conditions raises concerns about privacy violations and exposure to sensitive personal information.

The security challenges of AIoT healthcare systems are many and involve various aspects that must be taken into consideration, including technical, ethical, and legal dimensions. Healthcare providers and AIoT suppliers must implement adequate security plans and adhere to regulations and laws so that the healthcare industry can leverage the potential of AIoT healthcare effectively.

8.4 SECURITY SOLUTIONS AND STRATEGIES IN AIoT HEALTHCARE SYSTEMS

Overcoming AIoT-related vulnerabilities in healthcare systems requires the implementation of multiple strategies, including but not limited to solid data encryption, AI for threat detection and response, blockchain, federated learning, and Zero Trust Architecture (ZTA). This section examines these topics and technologies, focusing on the uses, challenges, and technical architectures that support them in order to suggest improvements at the cybersecurity level of healthcare systems.

Statistics indicate that the COVID-19 pandemic, ongoing hospital staff shortages, and greater reliance on interconnected devices have significantly altered the healthcare landscape. Electronic Health Records (EHR) and AIoT systems have experienced explosive growth that has changed the nature of the industry's positioning in the overall cyberattack threat matrix. In a report by IBM in 2023, the average amount of damage caused by data breaches in the healthcare sector is the largest at $10.93 million [32]. This strongly indicates a need for effective security measures to be implemented.

8.4.1 Encryption methods

Encryption protects sensitive information from unauthorized users by converting it into unreadable formats for users without access. Two encryption standards that are primarily used in healthcare systems are the Advanced Encryption Standard (AES) and Rivest-Shamir-Adleman (RSA).

AES: AES is a symmetric key-encryption technology adopted by The National Institute of Standards and Technology (NIST). The most robust AES variant, AES 256, has exceptionally high levels of security, because it performs 14 cycles of encryption in the form of expansion, permutation, and substitution [33]. AES can be used in many ways, such as encrypting and protecting electronic health records, ensuring secure communication between medical devices, and protecting data in cloud systems. AES-256 begins by adding a round key and then rotating between scales that account for several times. AES algorithms commonly appear in security libraries, such as OpenSSL and PyCrypto, for efficient utilization in healthcare applications.

- **Strengths**: Efficient strategies requiring long key lengths to resist brute-force attacks.
- **Limitations**: The concept of key management is challenging in a distributed environment, and improper implementation may lead to weaknesses and vulnerabilities.

Rivest-Shamir-Adleman (RSA) is an asymmetric algorithm that has gained popularity owing to its simple concept. The RSA uses a pair of keys: a public key for encrypting the message and a private key for decryption. Today, it is incorporated into many communication systems and is also helpful for sending encrypted data exchanges and authentication in the form of independent digital signatures within healthcare information systems [34]. RSA relies on the hardness of factorizing large composites or polynomial degree numbers to generate keys [42].

- **Strengths**: It is highly beneficial for providing SSL certificate-based security and secure communication.
- **Limitations**: The RSA is not suitable for encrypting large-scale data because of its computational complexity. RSA and similar algorithms are unlikely to withstand quantum computing because many quantum algorithms, such as Shor's, can break RSA.

8.4.2 Blockchain for Secure Data Sharing

Healthcare data can be managed and exchanged securely using blockchain technology as it is decentralized and resilient to fraudulent approaches. Transactions are secured within unchanged blocks, thus forming an unbroken chain of evidence [35].

Applications in healthcare:

- Smart contracts: Facilitate automated data exchange upon attaining predetermined conditions and legal requirements involving patient and data-sharing organizations [36].
- Patient's rights to have their data: Having private keys enables patients to allow or disallow access to their health files [36, 37].

Advantages:

- Improve the integrity of the data compared to conventional methods through decentralization of data storage, thereby minimizing the risks of data breaches.
- Open-source datasets and local infrastructure networks significantly accelerate and simplify the process of model development, genomic data, and other medical data for patients located at different institutions.

Challenges:

- Large medical datasets, for example, images, lead to scalability challenges.
- Migrating from the old methods and technology to the new one is time-consuming and tedious in most cases [37].

8.4.3 Federated learning

Federated learning (FL) is a form of distributed machine learning in which the participating nodes build the machine learning model locally without sharing central data. In this way, the data stay within the organizations, while the model is enhanced through collaboration [38]. An example of FL is FedAvg, a global federated model that receives only the aggregated parameters of a model trained locally, meaning that no local data are shared. Differential Data Protection techniques cover the individual contribution levels to provide more protection [39].

Applications in healthcare:

- Joint efforts can be made to train disease-predictive diagnostic models using the local data available to hospitals and research institutions.
- The deployment of FL is enhanced by frameworks, such as TensorFlow Federated or PySyft.

Advantages:

- Patient and sensitive data need not be shared.
- Complying with data sovereignty regulations such as the GDPR and HIPAA.

Challenges:

- The communication overhead caused during the model-updating processes can delay or prolong training time.
- Malicious nodes can strategize harmful local models, allowing them to exploit their vulnerabilities [39].

8.4.4 Zero Trust Architecture (ZTA)

The ZTA asserts that no person or entity within or outside the network can be trusted. This approach involves thorough verification of users, devices, and strict access controls [40]. The NIST zero-trust model offers a blueprint for those who wish to implement ZTA on critical infrastructure, specifically in access management, oversight engagement, and policy modification, covering any aspect where appropriate [40].

Core principles:

- Micro-segmentation: This reduces the potential movement of threats across a broad network by dividing it into many small and secure segments.
- Identity and Access Management (IAM): The concept of IAM guarantees that only authorized people are users of restricted resources.

Applications in healthcare:

- It helps protect sensitive patient information from unwanted access in EHR systems.
- It controls connected medical devices and ensures only authorized access [41].

Challenges:

- The implementation of ZTE requires a complete change in network infrastructure.
- Network resources may become overloaded mainly because of the ongoing monitoring required to maintain security.

As systems continue to be integrated across the healthcare industry, safeguarding sensitive information must remain a top priority in cybersecurity measures. Utilizing AES-256 and RSA encryption schemes, AI-based anomaly detection technologies, blockchain technology, federated learning, and zero-trust architectural systems are vital ideas that can be deployed.

8.5 CASE STUDIES IN AIoT HEALTHCARE SECURITY

8.5.1 Real-world security incidents in AIoT healthcare security

In recent years, healthcare systems have become increasingly prone to cyberattacks, mainly driven by the combination of the Internet of Things (IoT) and Artificial Intelligence (AI) in medical devices. While these technical developments have significantly boosted healthcare services at different levels and in different areas, they have also introduced new dangers by increasing the cyber surface. This section presents real-world case studies of cybersecurity in AIoT-enabled healthcare environments, focusing on cyberattacks and protection in some AIoT-enabled devices, such as insulin pumps and pacemakers [43].

8.5.1.1 Ransomware strikes on healthcare sector

Ransomware attacks have emerged as one of healthcare's most substantial cybersecurity dangers. These attacks include harmful software applications made to encrypt a victim's data, making them unattainable until a ransom is paid. Healthcare companies, because of their reliance on real-time data for personal care, are especially prone to such strikes, which can interrupt operations and endanger client security [44].

CASE STUDY: WANNACRY RANSOMWARE ATTACK ON THE NHS (2017)

The WannaCry ransomware attack in May 2017 was an archetype of the devastating consequences of cyberattacks on healthcare systems. This global cyberattack affected over 200,000 computer systems in 150 nations, and the UK Kingdom's National Health Service (NHS) was just one of the most critically affected institutions. The ransomware manipulated a vulnerability in Microsoft Windows systems, and the attack caused a cancellation or delay of more than 19,000 visits and procedures, and the interruption of patient treatment across various hospitals and facilities [45].

Cause: As the WannaCry attack illustrated, medical devices that did not receive critical updates issued by Microsoft were vulnerable to cyber threats and were infected.

Consequences: In addition to the chaos caused by healthcare service delivery, information was obtained regarding the possible vulnerabilities of IoT devices, which are either poorly configured or run on obsolete software that is not consistently patched and updated. The incident emphasized the importance of employing appropriate cybersecurity measures to protect healthcare systems against highly sophisticated threats [45].

8.5.1.2 Security breaches in IoT medical devices

IoT devices, such as insulin pumps, pacemakers, and other medical equipment, are incorporated into health facility systems to deliver patient services. However, the security of these devices is often compromised. Therefore, these devices compromise the security of more extensive networks and other systems, thereby becoming attractive targets for cybercriminals. The two following examples demonstrate weaknesses in the IoT medical devices [46].

CASE STUDY 1 VULNERABILITY IN AN INSULIN PUMP

In 2018, researchers found a severe flaw in some insulin pumps that could allow attackers to remotely control the devices and cause harm to diabetic patients from insulin overdoses. Researchers showed that by taking advantage of vulnerabilities in the wireless communication stack (e.g., Bluetooth) that most pumps utilize, they could wirelessly and remotely administer incorrect insulin doses to their target patients [47].

Cause: This weakness arose due to the poor security protocols employed in the communication channels of the insulin pumps. Several of these devices lack any form of encryption for their wireless communications, making them prone to attack.

Consequences: Even though vulnerability was not exploited during the research, such unsecured medical devices pose a severe potential threat. Consequently, manufacturers have been requested to implement higher security measures, including end-to-end encryption and multi-factor authentication, to ensure that critical medical devices are not procured in an unauthorized manner.

CASE STUDY 2 PACEMAKER VULNERABILITY

In 2017, the U.S. Food and Drug Administration (FDA) has noticed that some pacemakers made by Abbott Laboratories contain issues that make them potentially insecure. Pacemakers are intended to monitor heart rhythm, but abuse of the communication system was proven to be a weakness in which attackers could take control of the device [48].

Cause: The vulnerability was in the pacemaker's wireless communication system, which uses radio frequency technology to send data to external monitoring devices. Attackers can exploit this vulnerability by delivering inaccurate electrical pulses, thereby interfering with patient heartbeats.

Consequences: The breach highlighted the severe risks of wirelessly communicating with medical devices. It also re-evaluated the cybersecurity standards for medical tool manufacturers. In response, Abbott presented firmware updates and carried out more powerful protection measures, including file encryption and authentication attributes, to alleviate the risk of future attacks.

8.5.1.3 Statistics on healthcare cybersecurity breaches

The aforementioned threats belong to a larger picture of the constant growth of cybersecurity violations in the healthcare sector. According to a survey undertaken in 2020 by the Healthcare Information and Management Systems Society (HIMSS), cyberattacks in the healthcare industry increased by 74% from 2019 to 2020. In addition, the 2020 Verizon Data Breach Investigations Report stated that over 30% of all breaches pertained to healthcare institutions, where healthcare entities remained a highly targeted data breach sector [49].

Rise in ransomware attacks: The HIMSS report claims that nearly 50% of all attacks on the healthcare industry in 2020 were due to ransomware attacks. A notable aspect is that a direct increase was observed in incidents involving AIoT medical devices.

AIoT and data loss: IBM's X-Force conducted a survey and reported that in many instances, IoT devices used in the healthcare environment were the first to be penetrated by cyberattacks. The inadequate security mechanisms of the devices and unsupervised scanning of the networks make them susceptible to abuse.

8.5.1.4 Key takeaways and recommendations

These incidents signify the importance of improving cybersecurity in AIoT-driven healthcare systems. The NHS was compromised because of the WannaCry attack; there were some vulnerabilities in insulin pumps and security breaches in pacemakers. This indicates the need to understand the security threats posed by IoTs in healthcare. To mitigate these risks, healthcare organizations need to [50]:

1. Increase patch management: One known factor in fixing these issues is updating devices and systems with the latest security patches.
2. Enhancing device security: Companies that manufacture medical devices shall use encryption and secure communications technologies in the design of medical devices.

3. Implement Anomaly Detectors: Continuous surveillance and anomaly detection should help identify suspicious actions early and prevent breaches.
4. Healthcare Personnel: Healthcare personnel should be taught to be alert to potential cyber threats, the importance of securing patient data, and compliance requirements for devices powered by the IoT technology.

8.5.2 Successful implementations of AIoT security in healthcare

In this section, we concentrate on successful experiences of executing AIoT security within healthcare institutions, highlighting the strategies employed by organizations to ensure adequate protection of their networks, devices, and patients' data [51].

CASE STUDY 1 ADVANCED AI-DRIVEN SECURITY MONITORING AT MAYO CLINIC

The Mayo Clinic, which is widely recognized as one of the foremost healthcare facilities in the United States, has progressed immensely with the incorporation of AIoT technology while at the same time maintaining high cybersecurity standards. Thousands of connected medical devices are actively used in the operations of their hospitals and clinics. Therefore, the Mayo Clinic employs a defense-in-depth policy against cyberattack. Fulfilling this policy involves AI-based monitoring systems for medical devices designed to spot irregularities and initiate countermeasures in seconds [52].

Strategy employed: The Mayo Clinic implemented an AI-based anomaly detection system that continuously analyzes network traffic patterns and the activity of devices connected to the network. With machine-learning algorithms, the system can identify strange activities that can disclose a security breach, such as unauthorized access or the presence of malware. Apart from AI-based monitoring, the Mayo Clinic also employs other security features, including firewalls, intrusion detection system IDs, and MFA, to create a multi-layered approach to security management, where second layers of security still exist as backups when one layer is successfully compromised.

CASE STUDY 2 IOT DEVICE SECURITY AND RISK MITIGATION AT CLEVELAND CLINIC

The Cleveland Clinic is recognized for its critical medical services that utilize connected medical devices such as pacemakers, infusion pumps, and diagnostic tools. Extensive security measures have been implemented across the networks of these devices. Numerous vulnerabilities exist in IoT devices, highlighting the Cleveland Clinic's focus on safeguarding its medical devices while ensuring the confidentiality of patient data [53].

Strategy employed: The Cleveland Clinic employs a dual strategy of segmenting patient data networks and securing endpoints for IoT devices to prevent external attacks. If all devices are contained within one zone, the likelihood of penetrating critical networks is expedited. The advent of AI has helped in a new era, in which predictions are generated using reliable data streams from connected devices. These systems are designed to assess abnormal behavior and identify potential indicators of security breach. In addition, this type of AI detects and predicts unauthorized activities in advance.

CASE STUDY 3 COMPREHENSIVE SECURITY FRAMEWORK AT JOHNS HOPKINS MEDICINE

Johns Hopkins Medicine has established a strategic framework built to sustain research and excellence in healthcare, and it was complemented with an applied AIoT security architecture that integrates organizational and technical components. This security architecture aims to contain threats to AIoT technologies from internal and external sources while preserving the potential of AIoT technologies to deliver positive patient benefits [54].

Strategy employed: The implemented security framework by Johns Hopkins elaborated a comprehensive security architecture encompassing security measures, policies, and a set of tools comprising procedures to mitigate vulnerabilities at any level of the healthcare network. In addition, they deployed AI algorithms to forecast threats before an attack. The system continuously collects new data to identify hazards that may arise based on the trends and patterns observed in earlier attacks across the healthcare ecosystem. Regarding data storage and transmission, the Johns Hopkins security framework ensures that blockchain technology is utilized when dealing with sensitive data, which makes AIoT deployments more secure in the healthcare sector, ensuring that every data transfer is authenticated and that no data modification is possible [55]. Enhanced security of patient information, improved security of devices, and reduced security incidents are some of the benefits derived from the strategy employed by Johns Hopkins Medicine in AIoT security. Organizations have rapidly improved their ability to anticipate emerging threats through AI.

8.5.2.1 Key takeaways from successful implementations

Such cases demonstrate the advantages of distributed multi-layered security solutions in AIoT healthcare. The commonality in all of these successful operations is the implementation of AI or other technologies that assist in constantly protecting medical devices and health information. Cyber exposure significantly decreased in hospitals that could pursue an AI-enabled threat detection strategy, devise protection, and implement encryption, in addition to all other high security standards.

8.5.2.2 Recommendations for healthcare organizations

1. By learning from these experiences, organizations can utilize AI technologies to create safer and more resilient healthcare environments through the artificial IoT.
2. As the Internet of Things (IoT) advances healthcare, security measures evolve along with technological developments. These improvements aim to enhance the security and protection of medical device networks and sensitive patient data, ultimately contributing to a safe, effective, and sustainable healthcare system.
3. Healthcare organizations are to improve AI capabilities to support patient care and risk detection and implement proactive security measures to address emerging threats.
4. Healthcare organizations are to emphasize secure data sharing and utilize advanced technologies such as blockchain to ensure that patient information remains uncompromised during transit.

The experiences of significant healthcare providers should prompt other organizations to develop secure and sustainable methods for implementing AIoT healthcare environments.

8.6 FUTURE DIRECTIONS AND OPEN ISSUES IN AIoT HEALTHCARE SECURITY

Integrating AIoT technologies into healthcare has led to innovative approaches to existing processes. However, this advancement poses significant security risks that require further research and targeted solutions.

Additionally, concerns about the security of healthcare AIoTs emphasize the need to implement best practices, advanced technologies, and ethical standards in the future.

8.6.1 Quantum cryptography for secure communication

Given the perspective of quantum computing on encryption, some of the current encryption schemes in communications could no longer be safe to use. Quantum mechanics, particularly QKD principles, offers an entirely reliable security method. The security keys used in QKD make it virtually impossible for someone not authorized to monitor sensitive healthcare data encrypted using these keys. This technology can prevent unauthorized access to insulin pumps, pacemakers, and AIoT-related hardware from external sources [56].

8.6.2 Explainable AI (XAI) for transparency and accountability

AI systems in the healthcare industry often function as black-box systems, meaning that their decision-making processes are not easily understood. This concept is particularly relevant to explainable artificial intelligence (XAI). One of the primary goals of XAI is to enable scrutiny of the models used in healthcare settings. Understanding why an AI system identifies specific healthcare issues as risky is crucial for improving accountability and fostering trust in these technologies [57].

8.6.3 Standardization of AIoT security protocols

The use of different software, firmware, and communication protocols across various devices often hinders effective communication between IoT devices. Establishing a universally accepted standard for AIoT devices to communicate, encrypt sensitive information, and update firmware is crucial for enhancing user convenience and security. Additionally, protocols such as MQTT and ZigBee require improvements to incorporate robust encryption and authentication features, ensuring a standardized security framework for device interconnection [58].

One of the most significant reasons is the extensive range of security measures inherent in the broad concept of AIoT in healthcare. From a security standpoint, AIoT in healthcare encompasses everything, from simple wearable devices to complex diagnostic equipment. The absence of a unified standard for implementing AIoT devices results in numerous potential vulnerabilities that can be quite concerning [59]. Some challenges in creating universal standards for AIoT devices are as follows:

- Device heterogeneity arises from various manufacturers, firmware, and software stacks, which results in a wide range of AIoT devices. This diversity complicates the enforcement of uniform security standards across all devices [60].
- Interoperability standards enable AIoT devices to effectively communicate with healthcare IT systems. Without these standards, devices may function in isolation, thereby compromising their capabilities and security [60].
- It essential to address vulnerabilities because devices in a healthcare environment require continuous updates and patches. However, security gaps can arise when there are no standardized protocols for updating or patching IoT devices, particularly when dealing with devices from multiple manufacturers [60].

8.6.4 Addressing ethical and legal challenges

AIoT systems must adhere to several laws, including GDPR and HIPAA, while addressing moral issues, such as biases in AI models and the importance of patient confidentiality. In the future, the advancement of ethical AI will clarify who is accountable for inaccurate AI-driven translations that may harm patients. It is crucial to ensure that AI algorithm designs are free from biases and protect patient information through technological solutions, such as federated learning [61, 62].

8.6.5 Mitigating emerging threats

Adversarial AI algorithms, backdoor access to devices, and vulnerabilities in the supply chain can pose significant threats to the IoT networks. All AIoT security architectures must incorporate anomaly detection systems that quickly identify dangerous activities and tampering. Establishing strong partnerships between healthcare providers and cybersecurity specialists is essential to effectively combat these emerging threats [63]. Future AIoT healthcare security developments should include a multidisciplinary analysis of innovative technologies and standardization of protocols and regulations. To address existing challenges, AIoT could enhance health organizations by ensuring safety, privacy, and trust.

8.7 CONCLUSION

The artificial intelligence of things (AIoT) revolution in healthcare has significantly transformed the provision and management of medical services. With the Internet of Things (IoT) enabling real-time data collection and AI facilitating advanced analytics, healthcare systems can now deliver personalized care, improve diagnostic capabilities, and optimize operational efficiency. Practical applications of AIoTs, such as wearable sensors, remote monitoring devices, and AI-assisted surgical instruments, enhance patient care and health outcomes.

However, the integration of AIoT into healthcare systems presents several challenges. Security breaches, including attacks and vulnerabilities on medical devices, pose serious risks. These threats not only jeopardize patient safety, but also expose sensitive information and undermine the credibility of healthcare organizations. The financial losses and damage to reputation caused by such incidents are significant, but it is equally crucial to address the unique integration requirements of AIoT systems for healthcare providers.

A single operational framework is insufficient for addressing future challenges. However, technologies such as blockchain, federated learning, and zero-trust architecture are timely solutions for addressing security and privacy concerns in healthcare systems. Blockchain ensures secure data sharing, federated learning facilitates the training of artificial intelligence models while maintaining privacy, and zero-trust architecture mitigates risks by enforcing strict access policies. The development of explainable AI and quantum cryptography can enhance transparency and secure communication in AIoT systems.

Maximizing the potential of AIoT in the healthcare industry requires collaboration among all relevant stakeholders including care delivery organizations, regulatory authorities, and technology developers. Establishing standardized communication protocols and ensuring compliance with legal requirements, such as GDPR and HIPAA, is essential. Additionally, addressing ethical concerns related to AI decision-making is crucial. Furthermore, it is important to raise awareness of cybersecurity risks among healthcare professionals and foster a culture of vigilance to protect AIoT-based systems.

In conclusion, integrating AI and the Internet of Things (AIoT) presents transformative opportunities in healthcare. However, it is crucial to approach its implementation cautiously because it requires a balance between innovation and security. The healthcare sector can leverage AIoT to provide safe and environment-friendly services by addressing technical, ethical, and regulatory challenges. If implemented effectively, these technologies can enhance patient health while protecting patient safety and privacy.

REFERENCES

1. Zhang, J. and Tao, D. Empowering Things With Intelligence: A Survey of the Progress, Challenges, and Opportunities in Artificial Intelligence of Things. *IEEE Internet of Things Journal*, 2021. 8: p. 7789–7817.
2. Baker, S. and Xiang, W., Artificial Intelligence of Things for Smarter Healthcare: A Survey of Advancements, Challenges, and Opportunities. *IEEE Communications Surveys and Tutorials* 2023. 25: p. 1261–1293.

3. Sung, T. W., Tsai, P. W., Gaber, T., & Lee, C. Y., Artificial Intelligence of Things (AIoT) Technologies and Applications. Wireless Communications and Mobile Computing, 2021.
4. Shi, Q., Zhang, Z., Yang, Y., Shan, X., Salam, B., Lee, C., Artificial Intelligence of Things (AIoT) Enabled Floor Monitoring System for Smart Home Applications. *ACS Nano*, 15(11), 2021: p. 18312–18322.
5. Research, G.V., AI in Healthcare Market Size, Share & Trends Analysis Report by Technology, By Application, By End Use, And Segment Forecasts, 2022–2030. 2021.
6. Future, M.R., Healthcare IoT Market – Global Forecast to 2025. 2020.
7. Company, M., The COVID-19 Surge in Telemedicine Use: Insights and Opportunities. 2020.
8. Giuliano, K.K., Intravenous smart pumps: usability issues, intravenous medication administration error, and patient safety. *Critical Care Nursing Clinics of North America* 30(2), 2018: p. 215–224.
9. FDA, FDA Issues Recall of Infusion Pumps Due to Vulnerabilities. 2017.
10. Ghafur, S., et al., A retrospective impact analysis of the WannaCry cyberattack on the NHS. *NPJ Digital Medicine* 2(1), 2019: p. 98.
11. Team, AIME Planning. Artificial intelligence measurement and evaluation at the national institute of standards and technology. National Institute of Standards and Technology, 2021.
12. IBM Security. (2020). Cost of a data breach report 2020. IBM. Retrieved from https://www.ibm.com/security/data-breach
13. Greer, B.J., Cybersecurity For Healthcare Medical Devices. 2018, Utica College.
14. Singh, J., Sensor-Based Personal Data Collection in the Digital Age: Exploring Privacy Implications, AI-Driven Analytics, and Security Challenges in IoT and Wearable Devices. *Distributed Learning and Broad Applications in Scientific Research* 5, 2019: p. 785–809.
15. Naidu, G.A. and Kumar, J. Wireless protocols: Wi-Fi son, bluetooth, zigbee, z-wave, and Wi-Fi. in *Innovations in Electronics and Communication Engineering: Proceedings of the 7th ICIECE 2018*. 2019. Springer.
16. Peyroteo, M., et al., Remote monitoring systems for patients with chronic diseases in primary health care: systematic review. *JMIR mHealth and uHealth* 9(12), 2021: p. e28285.
17. Yousefirizi, F., et al., AI-based detection, classification and prediction/prognosis in medical imaging: towards radiophenomics. *PET Clinics* 17(1), 2022: p. 183–212.
18. Liu, Y., et al., Evolution of Surgical Robot Systems Enhanced by Artificial Intelligence: A Review. *Advanced Intelligent Systems* 6(5), 2024: p. 2300268.
19. Yu, M., et al., A posture recognition-based fall detection system for monitoring an elderly person in a smart home environment. IEEE Transactions on Information Technology in Biomedicine 16(6), 2012: p. 1274–1286.
20. Alharbi, H.A., Alharbi, K.K., and Hassan, C.A.U., Enhancing elderly fall detection through IoT-enabled smart flooring and AI for independent living sustainability. *Sustainability* 15(22), 2023: p. 15695.
21. Saxena, N., et al., Security and privacy issues in UK healthcare. Security and privacy of electronic healthcare records: concepts, paradigms and solutions, 2019: p. 283.

22. Quazi, F., Khanna, A., and Gorrepati, N. Data Security & Privacy in Healthcare. *Available at SSRN 4942328*, 2024.
23. Saxena, N., et al., Impact and key challenges of insider threats on organizations and critical businesses. *Electronics* 9(9), 2020: p. 1460.
24. Sarkar, K.R. Assessing insider threats to information security using technical, behavioural and organisational measures. *Information Security Technical Report* 15(3), 2010: p. 112–133.
25. Brass, I., et al. Emerging digital technologies in patient care: dealing with connected, intelligent medical device vulnerabilities and failures in the healthcare sector. in Workshop Report: PETRAS National Centre of Excellence in IoT Systems Cybersecurity. 2023. PETRSA & bsi.
26. Kalsi, J.S. and Ubhi, J.S. A study of Conventional Protocols applicable to the emerging IoT Systems and Devices. in *2019 International Conference on Automation, Computational and Technology Management (ICACTM)*. 2019. IEEE.
27. Dai, D. and Boroomand S. A review of artificial intelligence to enhance the security of big data systems: state-of-art, methodologies, applications, and challenges. *Archives of Computational Methods in Engineering* 29(2), 2022: p. 1291–1309.
28. Lee, H. and Lee, B. Addressing Emerging Threats: An Analysis of AI Adversarial Attacks and Security Implications. *International Journal of Advanced Smart Convergence* 13(2), 2024: p. 69–79.
29. Gao, Y., et al., Backdoor attacks and counter-measures on deep learning: A comprehensive review. *arXiv preprint arXiv:2007.10760*. 2020.
30. Bai, S., et al., Research on healthcare data sharing in the context of digital platforms considering the risks of data breaches. *Frontiers in Public Health*. 12, 2024: p. 1438579.
31. Scatiggio, V., Tackling the issue of bias in artificial intelligence to design ai-driven fair and inclusive service systems. How human biases are breaching into ai algorithms, with severe impacts on individuals and societies, and what designers can do to face this phenomenon and change for the better. 2020.
32. Mohsin, M., Salam, A.F., and Farokhnia Hamedani, M., Strategic Cybersecurity Management: The Impact of Knowledge Resources and Capabilities on Data Breach Risk. 2024.
33. Narasimha Rao, K.P. and Chinnaiyan, S., Blockchain-Powered Patient-Centric Access Control with MIDC AES-256 Encryption for Enhanced Healthcare Data Security. *Acta Informatica Pragensia* 13, no. 3 (2024): 374–394.
34. Nidhya, R., Shanthi, S., and Kumar, M. A novel encryption design for wireless body area network in remote healthcare system using enhanced RSA algorithm. in *Intelligent System Design: Proceedings of Intelligent System Design: INDIA 2019*. 2021. Springer.
35. Hussien, H.M., et al., Blockchain technology in the healthcare industry: Trends and opportunities. *Journal of Industrial Information Integration* 22, 2021: p. 100217.
36. Cyran, M.A., Blockchain as a foundation for sharing healthcare data. *Blockchain in Healthcare Today*, 2018.
37. Arbabi, M.S., et al., A survey on blockchain for healthcare: Challenges, benefits, and future directions. *IEEE Communications Surveys & Tutorials* 25(1), 2022: p. 386–424.

38. Bashir, A.K., et al., Federated learning for the healthcare metaverse: Concepts, applications, challenges, and future directions. *IEEE Internet of Things Journal* 10(24), 2023, 21873–21891.
39. Rauniyar, A., et al., Federated learning for medical applications: A taxonomy, current trends, challenges, and future research directions. *IEEE Internet of Things Journal* 11(5), 2023: pp. 7374–7398.
40. Edo, O.C., et al., A zero trust architecture for health information systems. *Health and Technology* 14(1), 2024: p. 189–199.
41. Syed, N.F., et al., Zero trust architecture (zta): A comprehensive survey. *IEEE Access* 10, 2022: p. 57143–57179.
42. Andreou, A., Mavromoustakis, C. X., Markakis, E. K., Mastorakis, G., Pallis, E., & Bourdena, A. Cryptography Solutions for Health Data. Intelligent Technologies for Healthcare Business Applications.
43. Vijarania, M., Gupta, S., Agrawal, A., & Misra, S., Achieving Sustainable Development Goals in Cyber Security Using AIoT for Healthcare Application. Springer Nature Switzerland, 2024: p. 207–231.
44. Bock, A., As Ransomware Attacks on Health Care Surge, Here's What Clinicians and Health Systems Can Do. *JAMA*, 2024. 332(8): p. 605–606.
45. Aljaidi, M., Ayoub, A., Samara, G., Alazaidah, R., Almatarneh, S., Khalid, M., Al-Gumaei, Y.A. *2022 International Engineering Conference on Electrical, Energy, and Artificial Intelligence (EICEEAI)*, 2022.
46. Martinez, J.B., *2018 9th IEEE Annual Ubiquitous Computing, Electronics & Mobile Communication Conference (UEMCON)*, 2018.
47. Yaqoob, T., Abbas, H. and Atiquzzaman, M. Security Vulnerabilities, Attacks, Counter-measures, and Regulations of Networked Medical Devices—A Review. *IEEE Communications Surveys & Tutorials* 21(4), 2019: p. 3723–3768.
48. Al-Juboori, S., Cyber-Securing Medical Devices Using Machine Learning: A Case Study of Pacemaker. *Journal of Informatics and Web Engineering*, 3, 2024: p. 271–289.
49. Seh, A.H., et al., Healthcare Data Breaches: Insights and Implications. *Healthcare* 8(2), 2020: p. 133.
50. Bradbury, S.L., Routledge Handbook of University-Community Partnerships in Planning Education. Key Takeaways. 2023.
51. Pise, A. A., Almuzaini, K. K., Ahanger, T. A., Farouk, A., Pant, K., Pareek, P. K., & Nuagah, S. J., Enabling Artificial Intelligence of Things (AIoT) Healthcare Architectures and Listing Security Issues. *Computational Intelligence and Neuroscience*, 2022, 8421434.
52. Adenekan, T.K., AI-Driven Diagnostic Models for Cardiovascular Health: Exploring Security and Business Analytics in Aortic Stenosis Detection. 2024.
53. Hempel, G.E., Do No Harm: Medical Device and Connected Hospital Security, in Women Securing the Future with TIPPSS for Connected Healthcare: Trust, Identity, Privacy, Protection, Safety, Security, F.D. Hudson, Editor. 2022, Springer International Publishing: Cham. p. 49–61.
54. Pronovost, P. J., Mathews, S. C., Chute, C. G., & Rosen, A. Creating a purpose-driven learning and improving health system: The Johns Hopkins Medicine quality and safety experience. *Learning Health Systems* 1 (1), 2017, e10018.
55. Jiang, Y., Xu, X., & Xiao, F. Attribute-Based Encryption With Blockchain Protection Scheme for Electronic Health Records. *IEEE Transactions on Network and Service Management*, 2022. 19: p. 3884–3889.

56. Shor, P.W. Algorithms for quantum computation: discrete logarithms and factoring. in *Proceedings 35th annual symposium on foundations of computer science*. 1994. Ieee.
57. Goodfellow, I., Deep learning. 2016, MIT press.
58. Lakshminarayana, S., et al., Securing the IoT Application Layer from an MQTT Protocol Perspective: Challenges and Research Prospects. *IEEE Communications Surveys & Tutorials*. 2024.
59. Andrii Stanko, O.D., Andrii Mykytyshyn, Oleg Totosko, Rostyslav Koroliuk, Artificial Intelligence of Things (AIoT): Integration Challenges and Security Issues. 2024.
60. Hussain, A., & Qureshi, K. N. (2024). Standards and Policies Adoption for AIoT Networks. In *Artificial Intelligence of Things (AIoT)* (pp. 49–76). CRC Press.
61. Act, Accountability Health insurance portability and accountability act of 1996. *Public Law* 104 (1996): p. 191.
62. Goddard, M. The EU General Data Protection Regulation (GDPR): European regulation that has a global impact. *International Journal of Market Research*. 59(6), 2017, p. 703–705.
63. Fleming, A.R., et al., Resilience and strengths of rural communities. Disability and vocational rehabilitation in rural settings: Challenges to service delivery. 2018: p. 117–136.

Chapter 9

AIoT security for healthcare

Enabling trust in smart medical devices

Areesha Fatima, Kishor Kumar Reddy C, and Monika Singh T

9.1 INTRODUCTION

Artificial Intelligence of Things integrates Artificial Intelligence and the Internet of Things, transforming the healthcare sector. AIoT represents a gateway for IoT devices, which connect with other devices sensibly, gather data, and provide resources and intelligence as well as analytics power, thus enabling systems to perform tasks automatically, improve operations, and deliver insights in real time [1]. When it comes to health applications, IoT is frequently referred to as the Internet of Medical Things (IoMT) or Healthcare IoT (H-IoT) [2]. Health care is where this convergence places innovation toward wearable health monitors, intelligent diagnostic devices, and robotic surgical assistants. AIoT has empowered service providers in healthcare with predictive analytics that enable them to predict diseases early, apply customized patient care plans, and improve patient outcomes. Fields such as remote patient monitoring, telemedicine, and routine automating tasks have impacted and relieved healthcare professional burden.

9.1.1 Role of AIoT in healthcare

The transformative role that AIoT will play in healthcare is that it brings together the powers of AI and IoT that can ultimately create intelligent systems that allow interconnected services. This, therefore, improves patient care, efficiency, and proactive health management. AIoT facilitates the collection of real-time data through wearable devices, smart monitors, and connected medical equipment, making it possible for healthcare providers to make informed decisions. It can predict health conditions based on the detection of pattern evolution and provide specific treatment plans [3]. IoT will enable seamless communication between appliances and platforms. This, in turn, makes remote patient monitoring, automated diagnostics, and precision robotic surgeries revolutionize the care delivery model by bringing early intervention as well as reducing hospital readmission. The AIoT not only streamlines the operational workflow but allows patients to take

 DOI: 10.1201/9781003606307-9

responsibility for their health through connected devices, thereby ushering in a collaborative style of well-being.

Smart health devices transformed the way modern healthcare actually works through the AIoT, using means to enhance the diagnostic scope and streamline patient care thereby enabling proactive management of health. Smart health devices seamlessly integrate IoT connectivity with AI's advanced data processing and analytical capabilities to make healthcare delivery more efficient, personalized, and patient-centric.

Smart medical devices and AIoT are advance modern healthcare systems. Medical devices are considered smart if they are connected. Glucose monitors, heart rate trackers, or AI-enabled imaging tools that allow for the constant flow of health data and insightful results. They create active health involvement among patients, as patients can engage in active management of their health through user-friendly interfaces and from a distance, access available medical support. For additional processing, the monitoring devices exchange the underlying health condition data gathered during the data-gathering phase. Other apps subsequently use the gathered data to track and manage various medical issues as needed [4]. These devices improve the accuracy of diagnosis and treatment and reduce hospital readmission rates, ensuring cost-effective delivery for healthcare providers. Smart medical appliances will enable chronic conditions to be monitored daily; because of this, healthcare systems will shift from reactive models of care to proactive models of care. Figure 9.1 illustrates the daily applications of AIoT in healthcare [1].

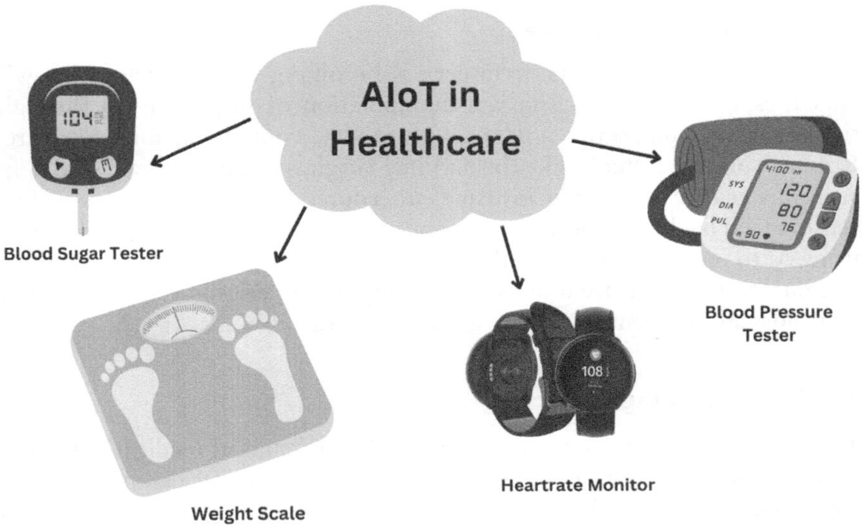

Figure 9.1 Applications of AIoT in healthcare.

9.1.2 The critical need for security and trust in AIoT-based healthcare

As for the healthcare domain, AIoT presents challenges in terms of security and trust. These smart medical devices hold sensitive information about patients, including PHI (Protected Health Information), making them a target for cyberattacks. Unauthorized access could expose the data breach, change the devices' functionalities, and bring forth risks to patient safety [1]. There is also an issue of choice in AI algorithms. Their problems of bias, clarity, and accountability need to be addressed. Wide adoption also goes hand in hand with trust in AIoT systems as the patient and healthcare providers alike have to be confident of its reliability, privacy, and ethical use. Thus, there is a need for secure measures such as encryption, identity verification, and compliance with the regulations in the healthcare sector.

9.2 AIoT APPLICATIONS IN HEALTHCARE

9.2.1 Remote patient monitoring (RPM)

Smart health devices transformed the way modern healthcare actually works through the AIoT, using means to enhance the diagnostic scope and streamline patient care, thereby enabling proactive management of health. Smart health devices seamlessly integrate IoT connectivity with AI's advanced data processing and analytical capabilities to make healthcare delivery more efficient, personalized, and patient-centric [5].

9.2.2 Robotic surgeries

The use of medical robots in surgery is becoming more widely acknowledged, especially for the accurate manipulation of surgical tools through tiny incisions under the guidance of robots, computers, and software. Robotic surgical systems, like the da Vinci Surgical System, utilize AI to help surgeons realize minimally invasive procedures with great precision. IoT sensors feed back into the system on the current condition of a patient while using AI to optimize the surgical pathway diminish errors and increase positive outcomes. These are best used for complex surgeries like any cardiac or neurological intervention, where accuracy is crucial [6].

9.2.3 Smart diagnostics

Medical imaging data account for up to 90% of clinical medical data, and intuitive and clear data is the key factor affecting clinical diagnosis. AIoT-enabled diagnostic tools, like intelligent imaging, can detect diseases with unprecedented accuracy by using machine learning algorithms in the power of IoT capabilities. For instance, an AIoT-powered radiology system can analyze

Table 9.1 Benefits of smart medical devices in healthcare

Benefit	*Description*	*Example*
Enhanced patient care	Improved diagnostics and personalized care	AI-assisted robotic surgeries
Real-time monitoring	Continuous tracking of patient vitals	Wearable heart monitors
Reduced hospital visits	Remote consultation and monitoring	Telehealth solutions
Faster treatment decisions	Data-driven insights for clinicians	AI-based diagnostic imaging

medical images including X-rays, MRIs, or CT scans to detect early signs of cancer, a fracture, and infection. In many cases, AIoT devices are much faster and more reliable than traditional methods [7]. Table 9.1 summarizes the advantages of AIoT-driven devices and their impact on healthcare delivery.

9.3 SECURITY CHALLENGES IN AIoT HEALTHCARE SYSTEMS

The integration of AIoT in the healthcare sector has resulted in unmatched benefits. However, it has also raised a lot of concern over security and trust. The smart medical devices collecting sensitive patient data, which are transmitted via telemedicine to various sources, become prime targets for cyberattacks. One of the biggest issues and worries with the IoMT system is maintaining the security and privacy of these data and records [8]. Data breaches may expose PHI, which further leads to identity theft, financial fraud, or misuse of medical data. Beyond data breaches, cyberattacks can immediately compromise patient safety; for example, an attacker manipulates the settings of a connected insulin pump or pacemaker and imperils patients' lives.

Moreover, the complexity of AI algorithm issues brings a transparency and accountability problem. Since many AIoT systems work like "black boxes," it becomes hard to understand decisions or check whether they are free from bias [9]. Lack of explainability often leads to mistrust among patients and healthcare providers, especially after a diagnostic error occurs or when AI predictions seem incongruent with clinical intuition.

This also makes the reliance on AIoT in health care a more contentious move in terms of data privacy issues. To start with, patients have to be assured that health-related information is kept and shared confidentially without unauthorized access and use. Compliance with current policy frameworks is automatic; however, how to comply with these standards within such diverse and interrelated AIoT systems is another gigantic task. Before being introduced to the market, clinical-grade medical devices need to be approved by the national regulatory body [10].

For establishing trust, security measures must be highly reliable and include end-to-end encryption, multi-factor authentication, and also device-specific firmware upgrades for securing their operation. Greater transparency through explainable AI frameworks can enable healthcare providers and patients to understand and validate AI-based decisions. Overall, the trust will require mechanisms in terms of accountability through audit trails and data provenance to ensure that AIoT systems are ethical and reliable.

Ultimately, solving these security and trust challenges is a matter of critical importance for realizing the proper promise of AIoT in healthcare. Without such robust protections and transparent operations, the adoption of these technologies will inevitably be stifled as they have little chance of delivering safer, smarter, and more efficient care.

9.3.1 Common vulnerabilities

Table 9.2 lists common vulnerabilities in AIoT systems, their impact, and examples from real-world incidents.

9.3.1.1 Device hacking

Smart medical devices are targeted by cyber attackers because these have important functions, and their in-built security features sometimes are minimalist. Attackers can exploit weak passwords, old software, and unpatched vulnerabilities, which would result in dangerous activities, such as tampering with the function of the medical device (like insulin pump dosages) or disabilities in its functionality, with direct impacts on patient safety [12].

9.3.1.2 Unsecured communication

The majority of AIoT devices rely upon an over-the-air communication protocol using Bluetooth, Wi-Fi, or cellular networks. If these links aren't

Table 9.2 Key security vulnerabilities in AIoT healthcare systems [11]

Vulnerability	*Impact*	*Real-world example*
Device hacking	Unauthorized access to medical devices	Hacking of insulin pumps
Insecure communication	Data interception during transmission	Ransomware attack on hospital networks
Firmware vulnerabilities	Exploitation of outdated software	Heartbleed bug affecting IoT devices
Insider threats	Misuse of access credentials	Employee leaking patient records

properly secured, they can become vulnerable to interception and tampering. Sensitive data is intercepted when transmitted between devices and servers without proper encryption. This allows the hacker to steal or manipulate that information [11].

9.3.1.3 Firmware attacks

Firmware is embedded software found controlling hardware devices and has become the most exploited by attackers. Via firmware modification or injection of malicious code techniques, the attacker may undermine device functionality or seize control persistently. In the healthcare context, this could mean that critical devices stop working or are repurposed to attack the network of the hospital [13].

9.3.2 Threats to data privacy, integrity, and availability in smart medical devices

9.3.2.1 Privacy breaches

Medical smart devices fetch massive amounts of PHI, which is why hackers look upon it as gold. It can lead to identity theft, insurance fraud, or even exposure to confidential and sensitive medical conditions, thus destroying the doctor-patient trust. Privacy and security concerns are a danger to user privacy and data confidentiality since unauthorized information storage is susceptible to integrity, privacy, and data security threats. The absence of a trustworthy authentication mechanism in IoMT devices makes this very evident [14].

9.3.2.2 Risks of data integrity

When all data values meet semantic requirements without being altered by unauthorized parties, this is referred to as data integrity [15]. Such attacks can lead to accuracy and reliability failure in a healthcare situation regarding patient information because virtual attacks that alter or destroy medical data could result in incorrect diagnoses or inappropriate treatments, putting at risk the lives of patients by misguiding the diagnostic results of a smart imaging device.

9.3.2.3 Availability disruption

There are incidents of ransomware and distributed denial-of-service attacks that are increasingly attacking healthcare systems. They can even lock out whole medical devices or hospital networks, thus further making critical treatments unavailable and the quality of care compromised. Such disruptions could be fatal in an emergency [10].

9.4 TRUST ISSUES IN AIoT-POWERED HEALTHCARE

9.4.1 Data misuse risks and unauthorized access

Unauthorized hackers, or even malicious insiders, could obtain such accessible information to steal, manipulate, or sell medical data. Apart from invasions of the confidentiality of patients, most such breaches can have extremely serious consequences, for example, identity theft, finance fraud, and even patient safety risks through manipulation of critical health data. In addition to these risks through external attacks, data misuse remains possible even by authorized system operators who handle patient data to achieve unapproved objectives like direct marketing or research in which permission was not sought from patients. Such risks point to the implementation of robust access controls, encryption, and auditing capabilities so that the security of data is ensured while also ensuring its ethical use [16].

9.4.2 Trust deficit between patients, healthcare providers, and technology

The integration of AIoT technologies was done very hastily in healthcare, and now it's seen to be growing as a deficit of trust among patients, service providers, and the technology itself. Patients express concerns over data privacy, the security of smart devices, and a fear of the oversight process being replaced by machines. The notion that biases exist in AI algorithms or their use increases this mistrust. On the other hand, healthcare providers may question the accuracy of insights generated from AI-driven systems if they cannot verify or understand how these systems work. Lack of trust in this way can hinder the full potential of AIoT from being tapped, even while it offers a lot of various benefits. A multiple approach would be needed to break such a trust deficit, including implementing open systems, compliance with privacy regulations, and invoking cooperation between the technology developers, providers, and patients. Trust must be built to exploit the possibilities of AIoT maximally to provide safe, efficient, and personalized healthcare.

9.5 SECURITY MEASURES FOR AIoT IN HEALTHCARE

Figure 9.2 illustrates the continuous cycle of managing security risks in IoT systems. The process involves identifying and understanding IoT assets, assessing associated risks, applying policies to reduce risks, preventing known threats, and detecting and responding to unknown threats to ensure robust IoT security.

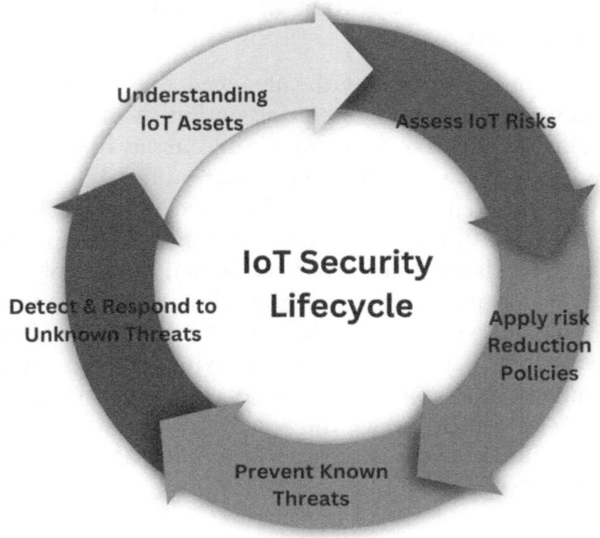

Figure 9.2 IoT security life cycle.

9.5.1 Data transition and at rest—end-to-end encryption

E2EE is one such key feature of secure data management in AIoT-based healthcare systems. Patient data from a wearable device would be encrypted right at the source with E2EE, meaning it would only decrypt at the destination, a designated server for example, or a cloud storage place [17]. For example, such data might be intercepted while in transit, but encryption would prevent sharing of the same with other players, hence still maintaining confidentiality. Also, encrypting data when it rests—the period it is stored on devices or servers—will prevent breaches due to theft or illegal access.

9.5.2 Identity and access management

Identity and access management is the best way to manage who gets to gain access to AIoT devices and the sensitive data they deal with [18]. Multi-factor authentication, for example, offers an added layer of security, wherein users will be required to provide several varieties of verification to gain access to systems such as a password, biometric scan, and so on. Role-based access control protects and ensures that people can only have access to the data and functionality required for their job roles. For example, a nurse can view the data of monitoring patients, but only a physician has the authority to change treatment plans. So IAM practices will diminish the risk of unauthorized access and insider threats and will enhance accountability through detailed log files recording every access.

9.5.3 Secure firmware updates and patch management

Firmware exists in all devices with advanced operation systems. Smart medical devices carry firmware operating within them that regulates their operations. This means that firmware should always be up to date with the latest updates. These updates help repair vulnerabilities and enhance functionality in different areas [19]. However, updating firmware itself needs to be secured to prevent an attacker from injecting malicious code. Secure firmware updates ensure that the actual source of the update is authenticated, for example, through digital signatures that updates are transmitted over encrypted channels [20]. Patch management also plays a big part because it is the process of identifying, testing, and distributing patches to known vulnerabilities in time. A comprehensive patch management strategy would then reduce the window for attackers, meaning that AIoT devices remain resilient against emerging threats while keeping their operational integrity.

9.6 TRUST-ENHANCING MECHANISMS

9.6.1 Explainable AI in support of transparency and accountability

Explainable AI or XAI is important to provide explainability support to AIoT-based healthcare systems so that they become more transparent and accountable. Explainable AI provides clear answers as to how decisions are being made, and it enables the users to understand their dependence on AI output and to build trust in such outputs [9]. For example, suppose an AI-based diagnostic tool shows that there might be a medical problem looming. In that case, XAI can detail the factors determining such a prediction, such as a trend in medical imaging or a rising trend in a patient's vital signs. The need for transparency is evident, and it is most acutely valued in the health sector where the decision point makes a difference in the patient's life. It ensures accountability and builds trust in the medical practitioner-patient relationship because XAI enables clinicians to check the AI's recommendations against possible errors or biases. This also meets the ethical and regulatory requirements because it can provide provable reasons for significant decisions.

9.6.2 Blockchain for secure and non-mutable medical records

Blockchain technology provides a secure, decentralized means for managing medical records in an AIoT system [21]. Blockchain further offers immutability of medical data by using distributed ledgers in such a way that, once recorded, it cannot be changed. Each transaction relating to the history of any patient is timestamped and cryptographically secured, which provides a fully transparent and verifiable audit trail; this capability is invaluable in assuring the integrity of sensitive health information and reducing the possibility of unwanted changes. Blockchains also facilitate safe data sharing among hospitals, insurance organizations, and research organizations while maintaining patient confidentiality. Patients can allow access to their data only with private keys to limit access only to authorized entities that view or make use of their information [22]. The integration of blockchain in healthcare will thus improve security in data protection, fewer cases of fraud, and interoperability [23].

9.6.3 Security standards and frameworks compliance

Creating a secure and trustworthy AIoT-based healthcare system would demand to follow security standards and frameworks. Regulations such as HIPAA, particularly for the United States, and GDPR for Europe mandate the measures of protection through encryption, access controls, and breach notification [3]. Industry-specific frameworks like ISO/IEC 27001—Information Security Management Systems—or NIST Cybersecurity Framework outline best practices for establishing security measures. Compliance ensures that healthcare organizations protect sensitive information but also develops trust among patients and stakeholders. Maintenance of compliance is challenging and never-ending because it demands regular auditing, employee training, and updates to security policies according to the changing nature of technologies and threats. Prioritization of compliance on the healthcare provider side will reduce risks, help in avoiding legal penalties, and will engender confidence in AIoT solutions.

9.7 CASE STUDIES AND APPLICATIONS

9.7.1 Real-world applications of AIoT security in healthcare

9.7.1.1 Medtronic's AIoT-powered remote monitoring system

Medtronic is one of the biggest medical device companies that have leveraged AIoT-driven remote monitoring systems for patients with chronic conditions such as diabetes and heart diseases. Their products, for instance, the MiniMed insulin pumps, cloud connect the patient's glucose levels and insulin delivery continuously [24]. In the case of Medtronic, it has taken a stance of securing all the data exchanged between the devices and healthcare providers by conducting full end-to-end encryption. Access to data about a patient for any healthcare professional comes with multi-factor authentication, hence increasing security. The secure infrastructure has enabled the healthcare provider to alter the treatment according to individual needs as per real-time data, thus helping improve patient outcomes and reducing the number of hospital visits.

9.7.1.2 IBM Watson health and AI for radiology

IBM Watson Health uses AI in the analysis of medical image data, which includes X-rays and MRIs, to diagnose diseases like cancer very early. The AI algorithms will process large volumes of imaging data while the IoT devices will stream this data between the hospital and IBM's cloud infrastructure, where end-to-end encryption and AI explainability tools ensure that there would be no such perversion or tampering in AI-driven diagnoses [25]. It also strictly follows rigorous compliance frameworks such as GDPR to protect patient data privacy. This has empowered clinicians to make proper and timely diagnoses that work for the patient's benefit.

9.7.2 Key take-aways

Table 9.3 summarizes notable breaches, their causes, and the lessons learned to improve security practices.

Preventive Cybersecurity Practices: Many healthcare organizations whose data was compromised had little or no good cybersecurity practices. A secure AIoT environment involves regular scanning for vulnerabilities, timely patch management, secure firmware updates, and continuous monitoring.

Risk Assessment End: This reduces risks to the fullest potential and ensures that AIoT systems are designed with security in mind from the design stage. This includes evaluating not just the devices themselves but the whole ecosystem: the cloud, data storage, and communication channels.

Patient Trust and Communication: Patient support for the data security policy of institutions is excellent if patients are educated on the appropriate measures that have been taken to safeguard their information. This information allows a higher level of trust from the patients and ensures that AIoT systems are not shunned in healthcare institutions.

Interdisciplinary Cooperation between Technology and Healthcare Experts: This will ensure that a group of cybersecurity experts, manufacturers of devices, and healthcare providers work together to identify and mitigate security risks effectively. In light of the above, a multidisciplinary

Table 9.3 Lessons learned from AIoT security breaches

Incident	*Cause*	*Lesson learned*
Ransomware attack on hospital network	Lack of patch management	Regular updates and vulnerability assessments
Data breach from wearable devices	Weak encryption protocols	Strong end-to-end encryption
Unauthorized access to patient records	Insider threats	Role-based access control and strict monitoring
Malware exploiting IoT device flaws	Unsecure firmware	Secure firmware updates and digital signature checks

approach will be very important in the development of secure and trustworthy AIoT systems in healthcare.

9.8 FUTURE DIRECTIONS

9.8.1 Emerging technologies to enhance security and trust

Table 9.4 highlights innovative technologies being applied to enhance security in AIoT healthcare systems.

9.8.1.1 Quantum cryptography

Quantum cryptography is an emerging technology based on the principles of quantum mechanics to produce virtually unbreakable encryption. As opposed to the class of encryption schemes that are possible with classical computation, quantum cryptography relies on principles in quantum superposition and entanglement as a way of protecting information during its transmission [26]. For instance, QKD can enable two parties to communicate such that if some eavesdropping happens, immediate detection will occur. Quantum cryptography can thus improve secure data transmission, preventing leakage of sensitive health data from AIoT healthcare systems across highly connected environments where man-in-the-middle attacks are fairly common. This technology has great promises to deliver the kind of security that is needed to protect personal health information in this increasingly digital, interconnected, and hyper-connected world.

9.8.1.2 Federated learning

Federated learning is an AI technique that makes it possible to train machine learning models jointly across multiple devices or institutions without even moving the actual data. In health care, patient data could be kept decentralized on local devices such as medical IoT or hospitals' servers with only the

Table 9.4 Emerging security technologies in IoMT

Technology	*Purpose*	*Example use case*
Quantum cryptography	Secure key distribution	Protecting remote patient monitoring data
Federated learning	Privacy-preserving AI model training	Decentralized analysis of health records
Blockchain	Immutable and transparent data storage	Managing electronic health records (EHRs)
Homomorphic encryption	Encrypted data processing	Secure cloud-based analytics

model updates being transmitted between the devices to improve the AI system. Federated learning contributes, therefore, to privacy by keeping one's healthcare data decentralized, a position that not only reduces possible risks of data breaches but also allows AI models to learn and update themselves. This will enhance trust because both patients and service providers will be assured that their data is not centrally stored or transferred to third parties, thus enhancing data privacy and security in AIoT healthcare systems [27].

9.8.2 Legislation and international cooperation in the advancement of secure AIoT systems

9.8.2.1 Legislation

Legislation plays a paramount role in defining the security aspects of AIoT in healthcare. There are also laws and standards that governments and regulatory bodies are creating regarding the safety, reliability, and respect for patient privacy of healthcare AIoT systems. For instance, HIPAA in the United States and GDPR in the European Union come under restrictions to be followed concerning the collection, processing, or storage of health data. Such regulations provide security requirements, such as encryption, access controls, and breach notification procedures. New laws specifically dedicated to AI, like the EU's Artificial Intelligence Act, are in place to influence regulation on the ethical use and security of AI technologies which requires aspects of transparency, accountability, and safety associated with healthcare systems powered by AI. With the increased adoption of AIoT, legislation will then become crucial in order to have standard security frameworks that eliminate all risks like data breaches, malicious attacks, and AI bias.

9.8.2.2 Global coordination

The continued complexity and connectivity of these AIoT systems in healthcare demand that their standards be set and frameworks organized through global coordination. With the increasing integration of AIoT technologies and healthcare systems across borders, socially shared awareness of practice and compliance standards is also required. International bodies, such as the WHO and ISO, continue working toward harmonization of guidelines and best practices to secure healthcare technologies. Since collaboration across the globe will bridge the gap of developing regions of varying technological maturity, all healthcare systems and organizational operations around the globe will be able to maintain robust security. This also offers an opportunity to exchange knowledge and best practices, giving healthcare practitioners from all over the world the opportunity to stay abreast of newer emerging threats and apply advanced security techniques for patient data, including quantum cryptography and federated learning. Table 9.5 provides an overview of key regulations shaping the security and privacy of AIoT systems in healthcare.

Table 9.5 Global regulations impacting AIoT in healthcare

Regulation/standard	*Region*	*Focus*	*Impact on AIoT systems*
HIPAA	United States	Data privacy and security	Encryption and access control requirements
GDPR	European Union	Personal data protection and consent	Strict data handling protocols
ISO/IEC 27001	Global	Information security management	Ensures end-to-end system security

9.9 CONCLUSION

AIoT-enabled healthcare systems indeed have the potential to be a paradigm shift for modern medicine regarding what patients can expect from their care, diagnoses, and remote monitoring. The effectiveness of AIoT-enabled healthcare systems relies on the ability to overcome the innate security vulnerabilities that come with interconnected devices and complex AI algorithms. End-to-end encryption and robust identity and access management, secure firmware updates, and blockchain-based recordkeeping must therefore be implemented in healthcare systems. By addressing these vulnerabilities, healthcare providers can minimize threats that include hacking into devices, data breaches, and unauthorized access, thereby ensuring patient information safety and privacy. Apart from these technology solutions, the promises of trustworthy AIoT systems in healthcare only rely on openness, technical conformity with global regulatory frameworks as observed in HIPAA and GDPR, and cooperation among and between industries and governments. The integration of explainable AI with other emerging technologies, including quantum cryptography and federated learning, further strengthens the resilience of the ecosystem. By learning from real-world implementations and adopting a proactive approach to risk management, the stakeholders can construct a secure, transparent, and patient-focused AIoT infrastructure. In the long run, these challenges will not only accelerate the adoption of AIoT in healthcare but also realize its full potential for revolutionizing medical care.

REFERENCES

[1] Pise, A. A., Almuzaini, K. K., Ahanger, T. A., Farouk, A., Pant, K., Pareek, P. K., & Nuagah, S. J. (2022). Enabling Artificial Intelligence of Things (AIoT) Healthcare Architectures and Listing Security Issues. *Computational Intelligence and Neuroscience*, 2022, 1–14. https://doi.org/10.1155/2022/8421434

[2] Baker, S., & Xiang, W. (2023). Artificial Intelligence of Things for Smarter Healthcare: A Survey of Advancements, Challenges, and Opportunities. *IEEE Communications Surveys & Tutorials*, 25(2), 1261–1293. https://doi.org/10.1109/COMST.2023.3256323
[3] Dwivedi, R., Mehrotra, D., & Chandra, S. (2022). Potential of Internet of Medical Things (IoMT) Applications in Building a Smart Healthcare System: A Systematic Review. *Journal of Oral Biology and Craniofacial Research*, 12(2), 302–318. https://doi.org/10.1016/j.jobcr.2021.11.010
[4] Taimoor, N., & Rehman, S. (2022). Reliable and Resilient AI and IoT-Based Personalised Healthcare Services: A Survey. *IEEE Access*, 10, 535–563. https://doi.org/10.1109/ACCESS.2021.3137364
[5] Ramalakshmi, K., Krishna Kumari, L., Rajalakshmi, R., & Theivanathan, G. (2024). Enhancing Healthcare Through Remote Patient Monitoring Using Internet of Things. In *Technologies for Sustainable Healthcare Development*, IGI Global (pp. 133–146). https://doi.org/10.4018/979-8-3693-2901-6.ch008
[6] Thacharodi, A., Singh, P., Meenatchi, R., Tawfeeq Ahmed, Z. H., Kumar, R. R. S. V. N., Kavish, S., Maqbool, M., & Hassan, S. (2024). Revolutionizing Healthcare and Medicine: The Impact of Modern Technologies for a Healthier Future—A Comprehensive Review. *Health Care Science*, 3(5), 329–349. https://doi.org/10.1002/hcs2.115
[7] Lu, Z., Qian, P., Bi, D., Ye, Z., He, X., Zhao, Y., Su, L., Li, S., & Zhu, Z. (2021). Application of AI and IoT in Clinical Medicine: Summary and Challenges. *Current Medical Science*, 41(6), 1134–1150. https://doi.org/10.1007/s11596-021-2486-z
[8] Zahedian Nezhad, M., Bojnordi, A. J. J., Mehraeen, M., Bagheri, R., & Rezazadeh, J. (2024). Securing the Future of IoT-Healthcare Systems: A Meta-Synthesis of Mandatory Security Requirements. *International Journal of Medical Informatics*, 185, 105379. https://doi.org/10.1016/j.ijmedinf.2024.105379
[9] Shin, D. (2021). The Effects of Explainability and Causability On Perception, Trust, and Acceptance: Implications for Explainable AI. *International Journal of Human-Computer Studies*, 146, 102551. https://doi.org/10.1016/j.ijhcs.2020.102551
[10] Wazid, M., Das, A. K., Mohd, N., & Park, Y. (2022). Healthcare 5.0 Security Framework: Applications, Issues and Future Research Directions. *IEEE Access*, 10, 129429–129442. https://doi.org/10.1109/ACCESS.2022.3228505
[11] Wazid, M., Das, A. K., Rodrigues, J. J. P. C., Shetty, S., & Park, Y. (2019). IoMT Malware Detection Approaches: Analysis and Research Challenges. *IEEE Access*, 7, 182459–182476. https://doi.org/10.1109/ACCESS.2019.2960412
[12] Biswas, A. R., & Giaffreda, R. (2014, March). IoT and cloud convergence: Opportunities and challenges. In *2014 IEEE World Forum on Internet of Things (WF-IoT)* (pp. 375–376). IEEE.
[13] Manikandan, J., & Choudhary, R. K. (2024). Fortifying Medical IoT Systems: A Comprehensive Analysis of Advanced AI Techniques for Network Security. *SSRN Electronic Journal*. https://doi.org/10.2139/ssrn.4999651
[14] Hasan, M. K., Ghazal, T. M., Saeed, R. A., Pandey, B., Gohel, H., Eshmawi, A. A., Abdel-Khalek, S., & Alkhassawneh, H. M. (2022). A Review on Security Threats, Vulnerabilities, and Counter Measures of 5G Enabled Internet-of-Medical-Things. *IET Communications*, 16(5), 421–432. https://doi.org/10.1049/cmu2.12301

[15] Sun, W., Cai, Z., Li, Y., Liu, F., Fang, S., & Wang, G. (2018). Security and Privacy in the Medical Internet of Things: A Review. *Security and Communication Networks*, 2018, 1–9. https://doi.org/10.1155/2018/5978636
[16] Hernandez-Jaimes, M. L., Martinez-Cruz, A., Ramírez-Gutiérrez, K. A., & Feregrino-Uribe, C. (2023). Artificial Intelligence for IoMT Security: A Review of Intrusion Detection Systems, Attacks, Datasets and Cloud–Fog–Edge Architectures. *Internet of Things*, 23, 100887. https://doi.org/10.1016/j.iot.2023.100887
[17] Haddad, A., Habaebi, M. H., Elsheikh, E. A. A., Islam, Md. R., Zabidi, S. A., & Suliman, F. E. M. (2024). E2EE Enhanced Patient-centric Blockchain-based System for EHR Management. *PLOS ONE*, 19(4), e0301371. https://doi.org/10.1371/journal.pone.0301371
[18] Awaisi, K. S., Hussain, S., Ahmed, M., Khan, A. A., & Ahmed, G. (2020). Leveraging IoT and Fog Computing in Healthcare Systems. *IEEE Internet of Things Magazine*, 3(2), 52–56. https://doi.org/10.1109/IOTM.0001.1900096
[19] el Jaouhari, S., & Bouvet, E. (2022). Secure Firmware Over-The-Air Updates for IoT: Survey, Challenges, and Discussions. *Internet of Things*, 18, 100508. https://doi.org/10.1016/j.iot.2022.100508
[20] Newaz, A. I., Sikder, A. K., Rahman, M. A., & Uluagac, A. S. (2021). A Survey on Security and Privacy Issues in Modern Healthcare Systems. *ACM Transactions on Computing for Healthcare*, 2(3), 1–44. https://doi.org/10.1145/3453176
[21] Rai, H. M., Shukla, K. K., Tightiz, L., & Padmanaban, S. (2024). Enhancing Data Security and Privacy in Energy Applications: Integrating IoT and Blockchain Technologies. *Heliyon*, 10(19). https://doi.org/10.1016/j.heliyon.2024.e38917
[22] Arora, P., & Makani, R. (2024). Blockchain Integration with AIoT Data Security and Privacy for Sustainability. *Journal of Artificial Intelligence and Systems*, 6(1), 112–123. https://doi.org/10.33969/AIS.2024060108
[23] Hussien, H. M., Yasin, S. M., Udzir, N. I., Ninggal, M. I. H., & Salman, S. (2021). Blockchain Technology in the Healthcare Industry: Trends and Opportunities. *Journal of Industrial Information Integration*, 22, 100217. https://doi.org/10.1016/j.jii.2021.100217
[24] Collyns, O. J., Meier, R. A., Betts, Z. L., Chan, D. S. H., Frampton, C., Frewen, C. M., Hewapathirana, N. M., Jones, S. D., Roy, A., Grosman, B., Kurtz, N., Shin, J., Vigersky, R. A., Wheeler, B. J., & de Bock, M. I. (2021). Improved Glycemic Outcomes With Medtronic MiniMed Advanced Hybrid Closed-Loop Delivery: Results From a Randomized Crossover Trial Comparing Automated Insulin Delivery With Predictive Low Glucose Suspend in People With Type 1 Diabetes. *Diabetes Care*, 44(4), 969–975. https://doi.org/10.2337/dc20-2250
[25] Park, T., Gu, P., Kim, C.-H., Kim, K. T., Chung, K. J., Kim, T. B., Jung, H., Yoon, S. J., & Oh, J. K. (2023). Artificial Intelligence in Urologic Oncology: The Actual Clinical Practice Results of IBM Watson for Oncology in South Korea. *Prostate International*, 11(4), 218–221. https://doi.org/10.1016/j.prnil.2023.09.001
[26] Chen, Y.-J., Hsu, C.-L., Lin, T.-W., & Lee, J.-S. (2024). Design and Evaluation of Device Authentication and Secure Communication System with PQC for AIoT Environments. *Electronics*, 13(8), 1575. https://doi.org/10.3390/electronics13081575
[27] Awaisi, K. S., Ye, Q., & Sampalli, S. (2024). A Survey of Industrial AIoT: Opportunities, Challenges, and Directions. *IEEE Access*, 12, 96946–96996. https://doi.org/10.1109/ACCESS.2024.3426279

Chapter 10

Securing healthcare with AIoT

Navigating the future of medicine in intelligent ecosystems

Danish Ali, Sundas Iqbal, Sumaira Rafique, and Fahad Rashad Khan

10.1 INTRODUCTION

The union of AI with the Internet of Things (IoT) that enables AI-based systems to optimize their behavior in real-time based on human inputs and environmental signals or changes is the essence of the Artificial Intelligence of Things (AIoT) [1]. Integrating AI with IoT is essential in different fields, and Internet of things for healthcare is one of them [2], for the utilization of massive amounts of patient data which are collected by smart devices, and processed by machine learning algorithms for aids and optimizing care delivery [3]. AIoT covers wearable devices, remote monitoring systems, diagnostic tools, and robotic surgery, all of which involve a network of interconnected devices with advanced AI processing for decision-making [4].

AIoT strongly influences patient care and operational effectiveness. Wearable devices allow healthcare providers to continuously monitor patients' vital signs so that he or she can turn to help patients when necessary, preventing a health crisis [5]. In addition, it enables intercommunication of medical equipment, allowing teams to monitor patients round the clock and automated alerts [6]. For example, hospitals can automate their administration tasks using AI-driven systems like scheduling and patient flow management and others such as inventory. Through AIoT, it is also feasible to get consultations online, and this reduces the need to physically visit a physician, and this makes healthcare accessible in geographical areas where there is the least availability of hospitals or clinics.

However, AIoT creates a lot of security risks, along with its unavoidable advantages. Sensitive patient data is collected and transmitted between cloud servers, creating risks associated with data breaches, unauthorized access, and cyberattacks [7]. It should be able to operationalize strict regulations like HIPAA (Health Insurance Portability and Accountability Act) and GDPR (General Data Protection Regulation) related to privacy and more. With any breach of the security of AIoT in healthcare, many lives can be lost, which means stricter security is required.

DOI: 10.1201/9781003606307-10

10.2 CHALLENGES OF AIoT SECURITY IN HEALTHCARE SYSTEMS

10.2.1 Unique security concerns in healthcare

The privacy of patient data is one of the most challenging concerns in the security of AIoT healthcare. Healthcare systems contain some of the most sensitive information that needs to be protected from any kind of unauthorized access, internal or external [8]. Moreover, AIoT devices that process sensitive health data from various stakeholders require stringent data protection measures to conform with health regulations (HIPAA, GDPR, etc.).

Awotunde et al. [9] have studied the privacy and security issues in IoT-based healthcare systems, as the integration of Internet of Things (IoT) technologies in healthcare environments is increasing. Abstract: The advent of wearable devices and wireless sensor networks has made continuous health monitoring possible, providing a more user-friendly and effective solution for improving healthcare and emergency response systems. Despite the benefits of IoT in healthcare, the authors emphasize the privacy and security threats in the transmission, processing, and storage of sensitive health data. Such risks include lack of access to medical information, data breaches, and health-threatening delays or compromises in treatment. To address these issues, the authors present a security architecture that will safeguard healthcare information while considering the privacy and security requirements of the healthcare industry and its associated IoT systems.

It explains how the increasing volumes of medical data are set to rise, thanks to the expansion of the Internet of Things (IoT), artificial intelligence (AI), and cloud computing technologies and the subsequent need for security protocols to appropriately care for the massive amounts of sensitive data generated. This chapter helps to understand the difficulties of providing adequate privacy and security in IoT-based healthcare systems and gives the solutions to start the process of mitigation. The framework proposed in this chapter is a starting point to tackle the very complicated security issues related to IoT healthcare settings but slightly more detailed case studies or evidence showing possible implementations could improve the work.

10.2.2 Interconnected devices and complexity

An AIoT system in healthcare generally consists of multiple inter-related devices with their unique security threats as shown in Figure 10.1. Medical devices such as pacemakers, infusion pumps, and other clinical devices may not have been designed with security standards in mind, making them vulnerable to later-exploitable security protocols. This problem becomes more severe with the growing use of IoT devices in hospitals and healthcare systems from multiple manufacturers, each under their individual security protocols [10]. Developing and executing a plan that mitigates the security of

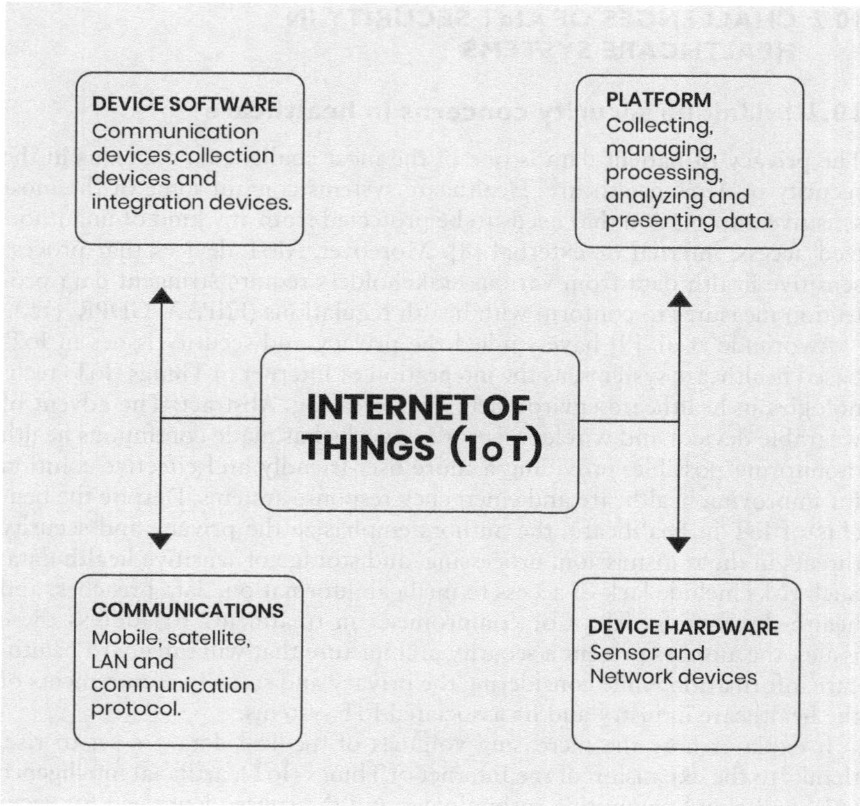

Figure 10.1 IoT and complexity of interconnected devices.

these devices is a complicated process that must also make sure that such devices are merged into a single integrated healthcare network within the system [11].

AL-mawee [12] addresses privacy and security issues in the context of Internet of Things healthcare applications for individuals with disabilities, focusing on Ambient Assisted Living (AAL) technologies which assist users to manage their daily life activities. The challenge that provides the context for the thesis is the aging population, and AAL, which draws on the diverse capabilities of the IoT, may play a crucial role in improving healthcare for the disabled. In this study, the authors first elucidate the fundamental privacy and security terminologies, which are then correlated and contextualized to healthcare IoT applications. This chapter reviews IoT architecture and components and then discusses how IoT can be used to fill some of the requirements of disabled users. The thesis highlights different applications of IoT with classification based on various kinds of disabilities. Afterward, privacy and security problems are analyzed, especially highlighting the specific threats to IoT-based healthcare systems for people with disabilities. It also surveys existing solutions that are IoT-based to combat these issues, providing a broad overview of the current solutions in the field. Finally, it highlights and outlines the privacy and security requirements needed for IoT health-related applications among disabled users which can act as a unique platform for further research and development. This study is a valuable addition to the debate on the use of IoT for the health of disabled people while remembering that privacy and security issues can be solved.

10.2.3 Cybersecurity threats and risks

Cyberattacks on healthcare systems have become more sophisticated, and ransomware, data theft, and Denial of Service (DoS) attacks are the most common types of risks [13]. The growing cyberattacks in the healthcare field has become a serious issue, a lot of which were directed at critical infrastructure during the COVID-19 pandemic in 2020. When a ransomware attack occurs in a healthcare organization, it affects patient care, can expose medical data, and delay critical treatments, all with potentially fatal results [14]. To protect AIoT healthcare environments, it is imperative to understand these threats and implement countermeasures as shown in Figure 10.2.

In a high growth sector of the future, the Internet of Medical Things (IoMT), Thomasian and Adashi [15] investigate cybersecurity issues associated with interconnecting hospital-based medical devices through the Internet to improve the delivery of patient care and operational efficiencies

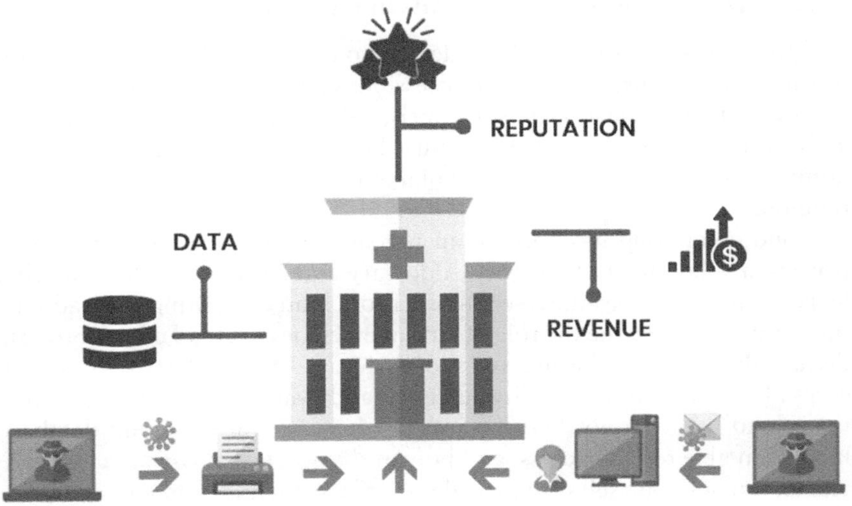

Figure 10.2 Flow of threats and risks in a healthcare system.

for hospitals. This hyperconnected environment poses unique challenges, and the chapter highlights the importance of proactive mitigation of cybersecurity risks to protect patient safety. The authors are specifically concerned with the United States, where they evaluate the policy tools currently in place to secure IoMT technologies. Using a qualitative approach, the study conducts a review of the literature including a comprehensive analysis of the federal and international legal documents, industry frameworks, and cyber breach analysis in order to highlight the relevant trends of regulation and the advantages and disadvantages of the current cybersecurity framework. Regulatory guidance has focused on areas including device identification, legacy devices, physical security, and breach detection, suggesting in more recent trends, an increasing desire at the federal level to enforce baseline security of devices. But the authors contend more regulatory guidance is needed for novel risks posed by retrofitted IT infrastructures, edge-to-cloud interfaces, and off-the-shelf device components.

Additionally, new cyber threats such as autonomous cyber-physical systems and quantum computing are only partially addressed by existing frameworks in this chapter. The authors emphasize that the integration of IoT into IoMT, when used together with a multistakeholder approach, can promote cyber hygiene and cybersecurity awareness. This can help improve the incident management processes and enhance the resilience of any cyber system operations [14]. Recognizing the state of regulation in this field, this review highlights existing gaps in the regulatory framework that will need to be addressed to protect and ensure the safety of IoMT-based healthcare technologies in the near future.

10.2.4 Data integrity and availability risks

Healthcare systems that use AIoT devices to provide real-time patient monitoring require trustworthy and accessible data [16]. Disruption or inaccuracy in data can lead to misdiagnosis, delayed treatment, or even death. This means the security of AIoT must be that data is both accurate and tamper-free and should also be available in front of doctors as and when required.

Cloud computing has become an urgent area of focus in data security, privacy, availability and integrity. Aldossary and Allen [4] explore the challenges that need to be addressed in order for cloud computing to be adopted on a larger scale, define the role of virtual machines in cloud computing, and discuss the resource sharing and cloud storage. The authors express concerns about data confidentiality, integrity, and availability by demonstrating that cloud users do not know whether other users in the cloud databases have been able to gain access, and how in the absence of trust in cloud providers, it would be useless to provide security authentication and authorization system provided by the cloud supplier to manage civil service security authentication and authorization. By conducting a survey on the existing solutions, this chapter discusses mitigation strategies for these risks, which are not presented in similar research, in relation to cloud computing in general. The authors detail the security and privacy technology adoption solutions that are being realized so that these areas can be more usefully part of the whole cloud computing process, and explore some remedial knowledge for users and self-governing groups alike, to know how cloud can be more secured.

10.3 AI-DRIVEN SOLUTIONS FOR ENHANCING HEALTHCARE SECURITY

10.3.1 Machine learning for threat detection

From identifying unusual trends of behavior, which may signify another security breach, at the same time greatly enhancing security in an AIoT-driven healthcare system, machine learning (ML) algorithms can do wonders. For example, ML can flag attempts to breach patient privacy and gain access to sensitive data or signal to administrators when unusual behavior is detected within medical devices. Machine Learning-based Intrusion detection systems help to monitor the system continuously and keep blocking the potential threats before they can damage the system [17]. For instance, there are AI-based security applications in hospitals that watch connected medical devices for unusual behavior and send alarms when they give indications that a security breach is about to occur.

Insider threat detection is a significant and recurring cybersecurity challenge, and it has been the subject of an extensive review by Farooq and Otaibi [18]. The broader study separated types of insider threats by type of employee, access level, motivation, and type of methods used, juxtaposed against actual incidents to agricultural a statistical analysis. The work highlights the machine learning techniques, detection methods, and evaluation metrics used to identify a malicious insider. While they acknowledge the progress, the authors note that biases in the studies already in the literature, the "missing real-world cases," and a scant emphasis on theoretical and technical components still present gaps. They describe also various challenges, threats to the value of related research, and implications for research in the field and provide recommendations on how the situation can be improved toward more effective insider threat detection systems. The share of the research dedicated to the usage of machine learning techniques, detection methods, and evaluation metrics for malicious insider detection are reported as follows. Even with progress, the authors cite biases in most studies, the absence of real-word cases, and little attention put toward the theoretical and technical issues, as limitations. They also analyze challenges in the field, provide recommendations to overcome challenges, and ultimately, develop more effective insider threat detectors.

10.3.2 Natural Language Processing (NLP) for data analysis

Natural Language Processing (NLP) can be utilized to offer advanced security for patient data and communications [19]. NLP techniques are able to process large amounts of textual information, such as patient records, clinical notes, or emails, for malicious patterns or suspicious activity [20]. For example, NLP might be drawn upon to determine the existence of phishing emails or fraudulent communications within a healthcare system. Further, NLP can protect the integrity of patient documentation by identifying any attempts to tamper with clinical data, as well as unauthorized data distribution in clinical situations [21].

Natural Language Processing (NLP) is another major area which became more relevant in the past years as the amount of textual data every day detonates by the home Internet usage of humans [13]. For example, the chapter highlights that NLP offers the opportunity to gain actionable insights into unstructured and unprocessed data domains, which will be key in settings like tourism where there is a large volume of free text that comes in the form of unlabeled data online. Utilizing machine learning approaches, NLP allows for more advanced text mining and provides researchers with the ability to better understand social phenomena and make more informed decisions. Submission authors also explain the text pre-processing procedures that are key for many useful NLP applications, and are important for understanding and implementing various NLP methods.

10.3.3 Predictive analytics for proactive security measures

Predictive analytics is a common tool in security space, and it can be leveraged within the healthcare field to manage threats ahead of time [8]. AI systems can predict possible security incidents based on historical data and trends. As an example, prescriptive models could discover tactics in attack maneuvers and predict the time and location of possible attacks, enabling healthcare organizations to take preventative measures. This can also reduce and alleviate the risks of security threats [9].

Adeniran et al. [3] discuss how predictive analytics technology is being used to significantly change security and risk management, and call to replace the traditional reactive system with one that is proactive. This chapter explores the theory behind predictive analytics and its relevance to practice across a range of domains, including cybersecurity, fraud prevention, and supply chain management. These include advantages such as support for early threat detection and increased allocation of resources, while some challenges related to data quality, privacy, and model explainability continue to be a problem. The authors further identify trends including artificial intelligence, real-time data analytics, and block chain as vital to adoption of predictive analytics in risk management. They end with the need to embed predictive analytics into the organization's risk management frameworks to help provide the level of resilience and adaptability changing complexity of risk demands. This chapter analyzes the theoretical foundations of predictive analytics and its implementation in spaces such as cybersecurity, fraud detection, and the supply chain analysis. Some of them help detect threats early and better allocate resources, but they still face challenges such as data quality, privacy concerns, and model interpretability. The authors suggest that emerging trends such as artificial intelligence, real-time data analytics, and block chain can all play key roles in furthering the predictive analytics continuum in risk management. Their belief that predictive analytics are integral to all aspects of risk management should be embedded within the framework that all organizations take to ensure resilience and adaptive capacity in a continually complex risk environment.

10.4 ADVANCED AIoT SECURITY TECHNOLOGIES AND PROTOCOLS

10.4.1 Blockchain for secure data sharing

Blockchain technology for secured data sharing is a newly emerging technology that securely connects device data to AIoT-enabled healthcare systems [22]. Blockchain preserves patient data in a tamper-proof manner and is accessible only to the authorized parties by decentralizing the storage. All transactions or updates to a patient health record are stored in an immutable ledger, which allows transparency and accountability [23]. Nevertheless, the application of blockchain in healthcare AIoT systems is not without its obstacles, including scalability issues and the requirement of interoperability with currently used systems.

Makhdoom et al. [21] propose "PrivySharing," a blockchain-based framework to enable privacy-preserving secure data sharing in smart city scenarios. The framework is designed to mitigate the risks of centralized IoT systems, with an emphasis on threats to availability, integrity, and privacy of data. To address this challenge, PrivySharing provides a multi-channel blockchain network, in which each channel processes only specific kind of data (e.g., health data, smart car data, financial data, etc.) and has a few pre-authorized institutions. Smart contracts enforce access control rules around channel data usage, and to further protect this information, private data collections of the shared data are encrypted. The client interaction layers integrate powerful API Key and OAuth 2.0 on both sides for strong security. Lastly, the framework also suggests the introduction of "PrivyCoin," a non-fungible digital token for incentivizing individuals to share data with stakeholders. In particular, PrivySharing meets specific needs of the European Union peering toward the General Data Protection Regulation (GDPR). We experimentally validate the scalability advantages of multi-channel blockchain as compared to single-channel ones and show that our approach is promising for IoT data sharing in smart cities.

10.4.2 Federated learning and edge AI for privacy-first approaches

Healthcare AIoT security and privacy-first solutions, powered by federated learning and edge AI, ensure data protection and efficient processing at the edge without compromising sensitive information. Federated learning, on the other hand, trains an AI model right on the original data instead of sending patient data to centralized servers for processing, eliminating the need to share any sensitive data with remote services and keeping it local and device-bound. Conversely, edge AI also allows real-time processing where data is collected, thus reducing inches during transmission where data can be captured [24]. They are of special importance for all mobile health devices and remote patient monitoring systems that must maintain patient privacy and process data instantly.

Li et al. [25] provide a comprehensive review of federated learning and its application in healthcare, emphasizing its potential to address data fragmentation and privacy concerns in medical AI. It also cites the absence of standardized electronic medical records, as well as the legal and ethical obligations to protect patients' privacy as specific challenges. A potential solution is federated learning, which allows AI models to be trained collaboratively without directly sharing raw data. In this work, the authors discuss its interaction with privacy-preserving algorithms, blockchain, and edge computing in terms of security and computational performance. The review describes an array of architectures, classification models, and healthcare applications of federated learning and elucidates its vulnerability toward various risks and attacks. Basic methods for privacy protection are discussed, as is the current status and limiting nature of federated learning implementation in medical settings. The chapter ends with an overview and outlook, highlighting the transformative characteristic of federated learning in enabling secure and efficient AI applications across the healthcare space.

10.4.3 Biometric and behavioral authentication

This has led to an increase in biometric authentication in healthcare, where patients and healthcare providers also utilize fingerprint scanning, facial recognition, and iris scanning to ensure that the individual in question is indeed who they claim to be, as shown in Figure 10.3. Another security layer in real time, because we could always analyze how the user is typing (it could be using patterns) and even the movement of the cursor (mouse) is a type of behavioral biometrics. By making it more difficult for unauthorized individuals to access sensitive healthcare data, these technologies help ensure

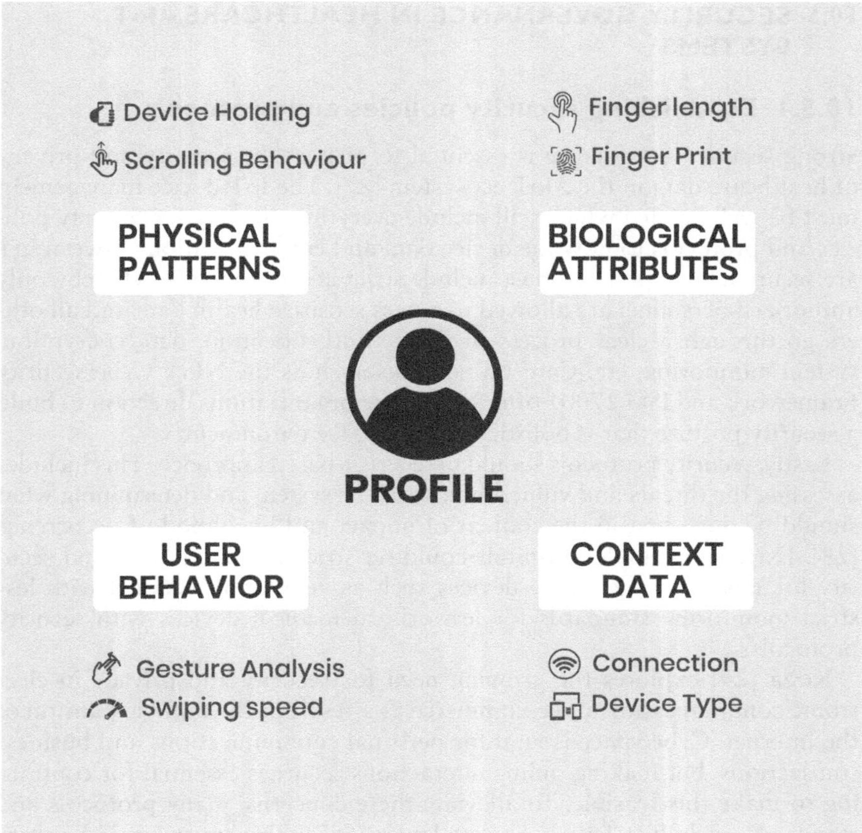

Figure 10.3 Biometric and behavioral authentication.

that only personnel with necessary permissions have access to this sensitive information [2].

Stylios et al. [26] provide an extensive survey on behavioral biometrics and continuous authentication technologies for mobile devices, offering valuable insights for researchers in this field. The chapter categorizes the different types of behavioral biometrics technologies and delves into methods involved in data collection and feature extraction. It contains a comprehensive literature review of state-of-the-art, highlighting how machine learning models perform over dimensions of seven types of behavioral biometric for continuous authentication. The survey also considers the susceptibility of these machine learning models to adversarial attack vectors and provides their countermeasures, where applicable. This work serves as a valuable guide for future research on secure and reliable user authentication methods on mobile platforms, with the authors completing their study with lessons learned, ongoing challenges, and future research directions.

10.5 SECURITY GOVERNANCE IN HEALTHCARE AIoT SYSTEMS

10.5.1 Establishing security policies and protocols

Strong security governance is essential to protect the security and privacy of healthcare data in the AIoT ecosystem [27]. The IoT device management must be well defined which will include everything from your security policies and protocols to how the device data and communication between IoT are secured. Such policies must include strict access controls, whereby only authorized personnel are allowed to access sensitive health data and all others go through a clear process of device authentication, data encryption, system monitoring, etc. Care frameworks such as the NIST Cybersecurity Framework and ISO 27001 offer healthcare organizations direction to build a security posture that is holistic for the AIoT environment.

Lastly, security protocols should take a risk-based approach. This includes assessing the threats and vulnerabilities in the system and determining what should be done first in the context of impact and likelihood of occurrence [28]. Extreme example: hospitals could set stricter standards around security for mission critical care devices such as ventilators coupled with less strict monitoring standards for less critical medical devices with security protocols.

Kizza [29] explores the growing need for security and privacy in electronic communication and e-commerce as a result of the rapid expansion of the Internet. Cyberspace is vital for personal communications and business transactions, but making online interactions secure is essential for continuing to make this feasible. To alleviate these concerns, many protocols and standards, including Secure Socket Layer (SSL), Transport Layer Security (TLS), secure IP (IPSec), Secure HTTP (S-HTTP), secure email protocols (such as PGP and S/MIME), DNSSEC, SSH, etc. This chapter looks at these protocols in detail, highlighting how terminologies are applied to maintain privacy and the importance of cryptography for the secure transmission of data across the Internet.

10.5.2 Risk management and compliance

Due to the complex regulatory landscape of healthcare, risk management must be a part of the established security governance framework [30]. Healthcare providers need to comply with local and international regulations HIPAA in the USA, GDPR in Europe, regional-specific privacy laws, and many others. Such legislation outlines specific expectations for the collection, processing, storage, and sharing of patient data. Not adhering to these regulations can calamitously affect a firm in terms of penalties and reputation, and senior health authority representatives can lose patient trust.

Regular security audits and vulnerability assessments are a fundamental part of risk management. Landoll [31] provides a means for finding weaknesses in the system before attackers can exploit them. Second, healthcare organizations also need an incident response policy that can respond effectively to a breach.

Van [32] examines the increasing adoption of machine learning in financial institutions (FIs) to enhance risk management and compliance processes, particularly through the use of regulatory technology (regtech). This includes use cases such as credit risk modeling, credit card fraud and money laundering detection, and the surveillance of conduct breaches, as outlined in the article. In other words, the study concludes two essential things about the value of machine learning to the finance sector. To begin, the impact of machine learning on analytical capabilities can be tremendous, especially with respect to money laundering detection and credit risk modeling, in situations where massive amounts of data must be processed and analyzed for deep, granular predictive insights. Second, machine learning is extremely context-dependent when it is used in financial services. It is not without some challenges on the quality and availability of data, and some machine learning models, being complex, are less interpretable and may lack explanatory power, particularly in a regulatory environment, since compliance teams need to understand and audit the model.

10.5.3 Incident response planning

Incident Response is an important component of AIoT security governance [33]. On the other hand, healthcare systems need to ready for efficient response actions against cyberattacks or security breach incidents to ensure that patient care and data quality are not impaired. An effective incident response plan will have processes for identifying the breach, containing the damage as soon as possible, and investigating the cause of the attack [34].

Clearly delineating roles and responsibilities for different response teams is essential. These teams should include experts from cybersecurity, healthcare administration, legal, and public relations. Health organizations also need to do frequent table top exercises to test preparedness and ensure that all stakeholders know how to act in the event of a breach. Transparent incident response requires the timely communication with the patients and regulatory bodies to ensure compliance [35].

Shinde and Priti [36] explain the digital landscape that is evolving rapidly and taking the world on the trends of cybersecurity. This piece freebie on the developing attributes of cyberattacks, both in their focus, and the strategies that are utilized. The authors call information theft the fastest-growing and most expensive cybercrime, with a steep acute trend over the past couple of years. Historically, criminal activity focused on the financial and private information central to organizational systems, but that has changed in recent years and threat vectors shifted toward industrial control systems. This transition is intended to interrupt industrial processes and wipe out vital data, creating new impediments for cyber safety defense and incident response. Given how tactics, techniques, and procedures in cyber always evolve, the paper calls for flexible approaches to response and planning for such things.

10.6 EXAMPLES AND REAL-WORLD APPLICATIONS

10.6.1 Example 1: AIoT security in hospital networks

A large hospital network in the United States implemented an AIoT security system to protect its interconnected medical devices and patient data. The hospital deployed a range of IoT devices, including smart infusion pumps, patient monitoring systems, and mobile health applications, all connected to a central network. However, the diversity of devices and their varying security standards posed significant challenges in managing security. To address these challenges, the hospital adopted a multi-layered security approach. This included integrating a blockchain-based patient data exchange system, securing the IoT network with machine learning-based anomaly detection, and using federated learning to train AI models locally on devices, without transmitting sensitive patient data. As a result, the hospital was able to significantly reduce the risk of cyberattacks and improve the overall security posture of its AIoT ecosystem.

10.6.2 Example 2: Remote patient monitoring systems and security

In a remote patient monitoring (RPM) program implemented by a leading healthcare provider in Europe, AIoT devices were used to continuously monitor patients with chronic conditions such as diabetes and heart disease. These devices collected real-time health data, which was transmitted to healthcare providers for analysis. Security concerns arose due to the need for transmitting sensitive patient data over potentially vulnerable networks. To mitigate these risks, the healthcare provider implemented end-to-end encryption and ensured secure transmission using virtual private networks (VPNs). In addition, the provider integrated anomaly detection algorithms to identify unusual patterns of data that could indicate potential security breaches. These measures not only ensured the privacy and integrity of patient data but also improved patient outcomes by providing real-time intervention.

10.6.3 Lessons learned and best practices

The key takeaway from these case studies is the importance of a layered security strategy. AIoT healthcare systems must integrate multiple security technologies, including encryption, machine learning-based threat detection, and decentralized data management, to protect against a range of potential threats. Additionally, continuous monitoring and real-time threat analysis are essential to quickly identify and mitigate security risks. Best practices include regularly updating security protocols, training healthcare personnel on security best practices, and ensuring compliance with relevant regulations. The summary of all studies included in the chapter is shown in Table 10.1.

10.7 FUTURE DIRECTIONS FOR AIoT SECURITY IN HEALTHCARE

The usage of AI and IoT in healthcare security will only evolve even beyond. Quantum encryption is one such area of great promise, as it can virtually guarantee security for sensitive health-related information. Quantum key distribution (QKD) allows communication channels that are invulnerable to cyberattacks, using traditional technologies; thus, there is a dawn of hope for securing patient data that is widely used for machine learning research. An interesting new domain is application of deep learning models for cyberattack prediction and prevention. Through the training of AI models on large datasets, systems can learn to detect behaviors of malicious activity, find vulnerabilities, and even respond autonomously to security incidents. These innovations will help healthcare providers be one step ahead of the next cyber threat and further strengthen AIoT healthcare networks.

One of the exciting developments we have around AIoT security is adaptive, self-healing systems. Such systems will be able to automatically identify and react to security threats without human intervention, facilitating immediate risk mitigation. Using machine learning and AI, adaptive systems can learn about new threats on an ongoing basis and change their security protocols. This method can help stomach the burden of cybersecurity professionals making sure as long some vulnerabilities adapt to enter their system, they are still protected. In healthcare, the effect of these systems would be tremendous. Self-healing security systems could provide safeguards by automatically detecting and suppressing potential threats for critical infrastructure such as medical devices and patient data. Such a level of automation can enhance the security and operational efficiency in health AIoT ecosystems.

Nonetheless, the future of healthcare is still being molded by AIoT technologies, and we need to ensure that ethical considerations regarding privacy, consent, and AI bias remain at the center of this evolution. Patients should be informed of all data that is collected and what it is used for. In addition, AI algorithms have to be created not to carry biases which lead to inequalities in healthcare results. AIoT systems must also adhere to confidentiality for patient(s) where personal data is anonymized whenever possible and with the patient's authority regarding access to their data.

Table 10.1 Summary of studies on AIoT security in healthcare: Objectives, methodologies, findings, and future directions

Author (year)	*Main objective*	*Methodology*	*Finding*	*Future work or recommendation*
Awotunde et al. [9]	Study privacy and security issues in IoT-based healthcare systems as IoT integration in healthcare increases.	Presented a security architecture to safeguard healthcare information, considering privacy and security requirements of healthcare IoT systems.	Identified difficulties in providing adequate privacy and security in IoT-based healthcare systems; proposed a security architecture as a starting point to tackle complex security issues in IoT healthcare settings.	More detailed case studies or evidence showing possible implementations could improve the work.
AL-mawee [12]	Address privacy and security issues in IoT healthcare applications for disabled users, focusing on Ambient Assisted Living (AAL) technologies.	Elucidated privacy and security terminologies, reviewed IoT architecture and components, discussed IoT applications for disabled users, analyzed privacy and security problems, surveyed existing IoT-based solutions, and outlined privacy and security requirements for IoT health-related applications among disabled users.	Reviewed IoT architecture and applications for disabled users; analyzed privacy and security problems; surveyed existing IoT-based solutions; highlighted and outlined privacy and security requirements needed for IoT health applications among disabled users.	Provided a platform for further research and development; emphasized that privacy and security issues can be solved.
Thomasian and Adashi [15]	Investigate cybersecurity issues associated with interconnecting hospital-based medical devices through the Internet to improve patient care and operational efficiencies.	Qualitative approach; literature review including analysis of federal and international legal documents, industry frameworks, and cyber breach analysis to highlight trends and regulatory gaps in IoMT security.	Highlighted the need for proactive mitigation of cybersecurity risks in IoMT; identified gaps in existing regulatory frameworks for novel risks like retrofitted IT infrastructures, edge-to-cloud interfaces, autonomous cyber-physical systems, and quantum computing.	Emphasized integration of IoT to IoMT, multi-stakeholder approach, promotion of cyber hygiene and cybersecurity awareness to improve incident management and resilience.

Aldossary and Allen [4]	Explore challenges in cloud computing regarding data security, privacy, availability, and integrity.	Conducted a survey on existing solutions related to cloud computing; discussed mitigation strategies for data security, privacy, availability, and integrity.	Highlighted issues with data confidentiality, integrity, and availability in cloud computing; discussed security and privacy technology adoption solutions; explored remedial knowledge for users and self-governing groups to enhance cloud security.	Presented security and privacy technology adoption solutions for cloud computing; explored remedial knowledge to enhance cloud security.
Farooq and Otaibi [18]	Review insider threat detection using machine learning techniques.	Extensive literature review; statistical analysis separating types of insider threats by employee type, access level, motivation, and methods used; compared against actual incidents.	Highlighted machine learning techniques, detection methods, evaluation metrics for malicious insider detection; identified biases, lack of real-world cases, and limited focus on theoretical and technical aspects in existing studies.	Provided recommendations to improve insider threat detection systems, overcome challenges, and develop more effective detectors.
Adeniran et al. [3]	Discuss how predictive analytics technology is used in security and risk management, advocating for proactive systems over reactive ones.	Explored theory behind predictive analytics; reviewed practice across domains including cybersecurity, fraud prevention, supply chain management; analyzed advantages and challenges.	Identified advantages of predictive analytics such as early threat detection and better resource allocation; challenges include data quality, privacy concerns, model explainability; highlighted trends like AI, real-time analytics, and blockchain as vital for adoption.	Recommend embedding predictive analytics into risk management frameworks to ensure resilience and adaptability in complex risk environments.

(Continued)

Table 10.1 (Continued)

Author (year)	*Main objective*	*Methodology*	*Finding*	*Future work or recommendation*
Usama Asim et al. (2022)	Propose "PrivySharing," a blockchain-based framework for privacy-preserving secure data sharing in smart city scenarios.	Designed a multi-channel blockchain network with smart contracts, encryption, integration of API Key and OAuth 2.0; introduced PrivyCoin as an incentive for data sharing; experimentally validated scalability advantages of multi-channel blockchain.	PrivySharing meets GDPR requirements; multi-channel blockchain offers scalability advantages over single-channel; promising for IoT data sharing in smart cities.	Potential application to smart city IoT data sharing; further scalability and implementation studies.
Li et al. [25]	Comprehensive review of federated learning and its application in healthcare, emphasizing its potential to address data fragmentation and privacy concerns.	Reviewed architectures, classification models, healthcare applications of federated learning; discussed privacy-preserving algorithms, blockchain, edge computing; analyzed vulnerabilities and risks.	Described various architectures and models of federated learning; elucidated vulnerabilities; discussed privacy protection methods; highlighted current status and limitations of federated learning in medical settings.	Highlighted transformative potential; suggested future outlook on federated learning for secure and efficient AI applications in healthcare.
Stylios et al. [26]	Survey on behavioral biometrics and continuous authentication technologies for mobile devices.	Categorized types of behavioral biometrics technologies; delved into data collection and feature extraction methods; comprehensive literature review of state-of-the-art; evaluated machine learning model performance over seven types of behavioral biometrics; considered susceptibility to adversarial attacks and countermeasures.	Provided insights into behavioral biometrics methods; highlighted susceptibility of machine learning models to adversarial attacks; discussed countermeasures.	Provided lessons learned, ongoing challenges, and future research directions for secure and reliable user authentication on mobile platforms.

Kizza [29]	Explore the growing need for security and privacy in electronic communication and e-commerce due to rapid expansion of the Internet.	Reviewed various security protocols and standards like SSL, TLS, IPSec, S-HTTP, PGP, S/MIME, DNSSEC, SSH; discussed application of terminologies and importance of cryptography for secure data transmission.	Highlighted the application of security protocols and standards to maintain privacy and security; emphasized the importance of cryptography in secure data transmission.	Not explicitly detailed
Van [32]	Examine the increasing adoption of machine learning in financial institutions to enhance risk management and compliance through regtech.	Discussed use cases like credit risk modeling, fraud detection, surveillance of conduct breaches; analyzed impact of machine learning on analytical capabilities; addressed challenges related to data quality, model interpretability.	Highlighted machine learning's impact on analytical capabilities in financial services; identified challenges like data quality, model interpretability in regulatory environments.	Recommended integrating machine learning within financial services while addressing data and model explainability challenges.
Shinde and Priti [36]	Explain the evolving digital landscape and trends in cybersecurity, focusing on cyberattacks and response strategies.	Discussed attributes of cyberattacks, strategies utilized, transitions in threat vectors; analyzed developing attributes of cyberattacks and defense strategies.	Highlighted that information theft is a fast-growing cybercrime; identified shift toward industrial control systems; emphasized need for flexible approaches to incident response and planning.	Called for flexible response and planning to address evolving cyber threats.

To overcome these risks from data misuse and at the same time enable the application of AI for better healthcare outcomes, privacy preserving technologies (like differential privacy and federated learning) can be leveraged.

10.8 CONCLUSION

AIoT in healthcare is changing the face of the healthcare industry by improving patient care, improving efficiency, and providing the ability to monitor patients in real time. This integration, however, brings large security dynamics. Partly due to the abundance of sensitive patient-related information and the need to maintain the efficiency of medical systems, healthcare organizations will need a data-protection-centric approach to security that accounts for everything from privacy days to the complexity of securing interconnected devices. Incorporating AIoT into healthcare systems requires the implementation of a new set of procedures and policies, which must be developed with greatest diligence. They need to be geared toward risk management and health regulatory compliance, and should utilize cutting-edge and next-generation technologies related to security, such as machine learning, blockchain, and federated learning. Routine security audits and planning for incident response are most definitely needed to keep a healthy and secure AIoT environment in the field of healthcare. Innovations in AI technologies, from quantum encryption to adaptive security systems, will create the future of AIoT security in healthcare. Together these innovations demonstrate a clear and applicable vision for how healthcare organizations can stay ahead of emerging threats and be best positioned to secure valuable patient data in an environment that grows increasingly complex and interconnected daily. Nonetheless, technology evolves and so will ethical and privacy concerns that will never lose its place as one of the most important considerations to ensure that AIoT can continue to serve the best interests of healthcare systems and patients' rights.

REFERENCES

[1] Abaoud M., Almuqrin M. A., and Khan M. F., "Advancing federated learning through novel mechanism for privacy preservation in healthcare applications," *IEEE Access*, vol. 11, pp. 83562–83579, 2023.
[2] Abouelmehdi K., Beni-Hessane A., and Khaloufi H., "Big healthcare data: preserving security and privacy," *Journal of Big Data*, vol. 5, no. 1, pp. 1–18, 2018.
[3] Adeniran I. A., Efunniyi C. P., Osundare O. S., and Abhulimen A. O., "Enhancing security and risk management with predictive analytics: A proactive approach," *International Journal of Management & Entrepreneurship Research*, vol. 6, no. 8, 2024, 32–40.
[4] Aldossary S. and Allen W., "Data security, privacy, availability and integrity in cloud computing: issues and current solutions," *International Journal of Advanced Computer Science and Applications*, vol. 7, no. 4, 2016, 485–498.

[5] Baker S. and Xiang W., "Artificial intelligence of things for smarter healthcare: A survey of advancements, challenges, and opportunities," *IEEE Communications Surveys & Tutorials*, vol. 25, no. 2, pp. 1261–1293, 2023.
[6] Areia C., King E., Ede J., Young L., Tarassenko L., Watkinson P., and Vollam S., "Experiences of current vital signs monitoring practices and views of wearable monitoring: A qualitative study in patients and nurses," *Journal of Advanced Nursing*, vol. 78, no. 3, pp. 810–822, 2022.
[7] Süzen A. A., "Cyber attacks for data breach and possible defense strategies in Internet of Healthcare Things ecosystem," *International Journal of 3D Printing Technologies and Digital Industry*, vol. 7, no. 1, pp. 55–63, 2023.
[8] Imoize A. L., Balas V. E., Solanki V. K., Lee C. C., and Obaidat M. S., Eds., *Handbook of Security and Privacy of AI-Enabled Healthcare Systems and Internet of Medical Things*, CRC Press, 2023.
[9] Awotunde J. B., Jimoh R. G., Folorunso S. O., Adeniyi E. A., Abiodun K. M., and Banjo O. O., "Privacy and security concerns in IoT-based healthcare systems," in *The Fusion of Internet of Things, Artificial Intelligence, and Cloud Computing in Health Care*, Springer International Publishing, Cham, 2021, pp. 105–134.
[10] Bhuiyan M. N., Rahman M. M., Billah M. M., and Saha D., "Internet of things (IoT): A review of its enabling technologies in healthcare applications, standards protocols, security, and market opportunities," *IEEE Internet of Things Journal*, vol. 8, no. 13, pp. 10474–10498, 2021.
[11] AlTawy R. and Youssef A. M., "Security tradeoffs in cyber physical systems: A case study survey on implantable medical devices," *IEEE Access*, vol. 4, pp. 959–979, 2016.
[12] AL-mawee W., "Privacy and security issues in IoT healthcare applications for the disabled users: A survey," 2012, *Masters Theses*.
[13] Egger R. and Gokce E., "Natural language processing (NLP): An introduction: making sense of textual data," in *Applied Data Science in Tourism: Interdisciplinary Approaches, Methodologies, and Applications*, Springer International Publishing, Cham, 2022, pp. 307–334.
[14] Farringer D. R., "Send us the bitcoin or patients will die: Addressing the risks of ransomware attacks on hospitals," *Seattle UL Rev.*, vol. 40, pp. 937–957, 2016.
[15] Thomasian N. M. and Adashi E. Y., "Cybersecurity in the internet of medical things," *Health Policy and Technology*, vol. 10, no. 3, p. 100549, 2021.
[16] Taimoor N. and Rehman S., "Reliable and resilient AI and IoT-based personalised healthcare services: A survey," *IEEE Access*, vol. 10, pp. 535–563, 2021.
[17] Chaabouni N., Mosbah M., Zemmari A., Sauvignac C., and Faruki P., "Network intrusion detection for IoT security based on learning techniques," *IEEE Communications Surveys & Tutorials*, vol. 21, no. 3, pp. 2671–2701, 2019.
[18] Farooq H. M. and Otaibi N. M., "Optimal machine learning algorithms for cyber threat detection," in *2018 UKSim-AMSS 20th International Conference on Computer Modelling and Simulation (UKSim)*, IEEE, 2018, pp. 32–37.
[19] P. N. K. Sarella and V. T. Mangam, "AI-driven natural language processing in healthcare: transforming patient-provider communication," *Indian Journal of Pharmacy Practice*, vol. 17, no. 1, pp. 21–26, 2024.
[20] Arjunan T., "Detecting anomalies and intrusions in unstructured cybersecurity data using natural language processing," *International Journal for Research in Applied Science and Engineering Technology*, vol. 12, no. 9, pp. 10–22214, 2024.

[21] Makhdoom I., Zhou I., Abolhasan M., Lipman J., and Ni W., "PrivySharing: A blockchain-based framework for privacy-preserving and secure data sharing in smart cities," *Computers & Security*, vol. 88, p. 101653, 2020.

[22] Rao S. S., Fernandes S. L., Singh C., and Gatti R. R., Eds., *AIoT and Big Data Analytics for Smart Healthcare Applications*, Bentham Science Publishers, 2023.

[23] Peltier T. R., *Information Security Policies, Procedures, and Standards: Guidelines for Effective Information Security Management*, CRC Press, 2016.

[24] Mohanna A., "Edge Artificial Intelligence for Real-Time Target Monitoring," 2023.

[25] Li H., Li C., Wang J., Yang A., Ma Z., Zhang Z., and Hua D., "Review on security of federated learning and its application in healthcare," *Future Generation Computer Systems*, vol. 144, pp. 271–290, 2023.

[26] Stylios I., Kokolakis S., Thanou O., and Chatzis S., "Behavioral biometrics & continuous user authentication on mobile devices: A survey," *Information Fusion*, vol. 66, pp. 76–99, 2021.

[27] Pise A. A., Almuzaini K. K., Ahanger T. A., Farouk A., Pant K., Pareek P. K., and Nuagah S. J., "Enabling artificial intelligence of things (AIoT) healthcare architectures and listing security issues," *Computational Intelligence and Neuroscience*, vol. 2022, no. 1, p. 8421434, 2022.

[28] Pfleeger C. P. and Pfleeger S. L., *Analyzing Computer Security: A Threat/Vulnerability/Countermeasure Approach*, Prentice Hall Professional, 2012.

[29] Kizza J. M., "Computer network security protocols," in *Guide to Computer Network Security*, Springer International Publishing, Cham, 2024, pp. 409–441.

[30] Ksibi S., Jaidi F., and Bouhoula A., "A comprehensive study of security and cyber-security risk management within e-Health systems: Synthesis, analysis and a novel quantified approach," *Mobile Networks and Applications*, vol. 28, no. 1, pp. 107–127, 2023.

[31] Landoll D., *The Security Risk Assessment Handbook: A Complete Guide for Performing Security Risk Assessments*, CRC Press, 2021.

[32] Van Liebergen B., "Machine learning: A revolution in risk management and compliance?" *Journal of Financial Transformation*, vol. 45, pp. 60–67, 2017.

[33] Reddy A. R. P. and Ayyadapu A. K. R., "Automating incident response: AI-driven approaches to cloud security incident management," *Chelonian Research Foundation*, vol. 15, no. 2, pp. 1–10, 2020.

[34] Ozkaya E., *Incident Response in the Age of Cloud: Techniques and Best Practices to Effectively Respond to Cybersecurity Incidents*, Packt Publishing Ltd., 2021.

[35] Ajmal, C. S., Yerram, Sravani, Abishek, V., et al. Innovative Approaches in Regulatory Affairs: Leveraging Artificial Intelligence and Machine Learning for Efficient Compliance and Decision-Making. *The AAPS Journal*, 2025, vol. 27, no 1, p. 22.

[36] Shinde N. and Kulkarni P., "Cyber incident response and planning: A flexible approach," *Computer Fraud & Security*, vol. 2021, no. 1, pp. 14–19, 2021.

Chapter 11

AI and IoT in smart healthcare

Transforming patient care and enhancing security

Hind Moussaid, Khawla Jabari, and Abderrahim Abdellaoui

11.1 INTRODUCTION

The healthcare sector faces unprecedented challenges, including the exponential growth in healthcare demands, limitations in medical resources, and the need to improve accessibility and personalization of treatments. To address these challenges, the integration of emerging technologies such as Artificial Intelligence (AI) and the Internet of Things (IoT) has become a priority. These technologies are transforming healthcare by offering innovative solutions to enhance diagnostics, treatments, and patient monitoring while addressing the growing demand for efficient, accessible, and secure care [1, 2].

Recent advancements in the application of AI and IoT in modern medicine are remarkable. For instance, an Accenture study highlights that 73% of global healthcare leaders have integrated AI into various operational processes, underlining its growing importance. AI-powered tools, such as radiological image analysis systems, are expected to be involved in 50% of clinical incidents by 2022, while virtual health assistants could manage up to 20% of patient interactions. Simultaneously, the global healthcare IoT market is projected to generate $108.60 billion in revenue by 2024, with a compound annual growth rate (CAGR) of 11.47% until 2028 [1], establishing itself as a critical technology in smart healthcare.

The Internet of Medical Things (IoMT) combines connected devices and advanced sensors to provide continuous patient health monitoring, enabling rapid and accurate clinical decision-making [3, 4]. Additionally, AI models such as Machine Learning (ML) and Deep Learning (DL) allow real-time analysis of vast medical datasets, improving diagnostics and predictive capabilities. These technologies hold transformative potential, particularly in personalized medicine and telehealth, optimizing care delivery while reducing costs.

However, adopting these innovations comes with significant challenges. Concerns related to data protection, ethical limitations, and technical integration remain prominent obstacles. For example, reliance on complex infrastructures such as cloud, fog [5, 6], and edge computing raises

DOI: 10.1201/9781003606307-11

questions about latency, network resilience, and the security of sensitive information [7]. Furthermore, a lack of understanding and trust in these technologies among healthcare professionals sometimes limits their widespread adoption.

This chapter explores the contributions of AI and IoT to smart healthcare, examining their advantages and practical applications while addressing the ethical, technical, and security challenges they present. It highlights potential strategies to navigate these challenges, emphasizing the importance of interdisciplinary collaboration and innovation. By situating these technologies within a smart healthcare framework, this review demonstrates their transformative potential to revolutionize the medical sector, paving the way for improved healthcare outcomes and operational efficiencies.

11.2 IOT AND HEALTHCARE

The integration of the Internet of Things (IoT) into healthcare, often referred to as the Internet of Medical Things (IoMT), has revolutionized the collection, analysis, and utilization of medical signals to enhance patient care. This section explores the diverse applications of IoT in healthcare, including patient monitoring, intelligent medical devices, and medical infrastructure management.

11.2.1 IoT, IoMT, and medical signals

IoT and IoMT technologies have significantly enhanced the collection and analysis of medical signals, enabling advanced diagnostic and monitoring capabilities. For instance, a study utilized a multi-sensor platform integrating single-channel ECG and dual-channel pressure pulse wave (PPW) inputs to measure blood pressure. Using a weakly supervised feature selection (WSF) method, 35 physiological features were extracted and reduced to identify subject-specific indicators, showcasing the potential for personalized health monitoring [8].

In another application, emotion [9] recognition systems based on IoT-enabled multi-sensor platforms combined EEG, EOG, and EMG signals with vital parameters like body temperature and respiration. These systems achieved accuracies of 72% and 89% on the DEAP and SEED datasets, respectively, through the integration of a Support Vector Machine (SVM) with sequential backward selection (ST-SBSSVM). This demonstrates the role of IoT in advancing mental health diagnostics.

Beyond healthcare, IoT-based systems have shown promise in critical environments such as mining. Gu et al. proposed a real-time monitoring system using Random Forest (RF) and SVM-based data fusion models to evaluate worker safety in hazardous conditions. The same principles of IoT data integration can be applied in emergency healthcare scenarios to improve real-time decision-making. These examples highlight the transformative potential of IoT in integrating complex physiological and environmental data for healthcare and safety-critical applications [10].

11.2.2 Patient monitoring

Remote patient monitoring (RPM) is one of the most transformative applications of IoT in healthcare, enabling continuous observation of patients' vital signs regardless of their location. IoT-enabled devices, such as connected glucometers, airflow monitors, and implanted cardiac devices, collect real-time physiological data. This data is automatically transmitted to secure databases and healthcare professionals for analysis. If abnormalities are detected, alerts are generated, allowing for timely and personalized medical interventions. Such systems not only improve patient outcomes but also reduce the need for prolonged hospital stays and prevent readmissions by identifying complications at an early stage.

The architecture of an RPM system, illustrated in Figure 11.1, demonstrates the seamless integration of wearable sensors, communication protocols, and fog/edge computing devices to ensure real-time data collection and processing. Wearable devices collect a variety of physiological signals, such as SpO_2, heart rate, blood glucose, temperature, and blood pressure, which are then transmitted via protocols like Zigbee, LoRa, and WiFi. The data is processed by fog/edge devices before being routed to healthcare providers for appropriate actions. This pipeline highlights the crucial role of IoT in bridging the gap between patients and healthcare professionals, particularly in remote settings [11].

An integral component of RPM is telemedicine, which leverages IoT technologies to provide remote healthcare services and enhance accessibility, particularly in underserved regions. Through secure IoT-enabled platforms, patients can transmit real-time health data—such as blood pressure, glucose

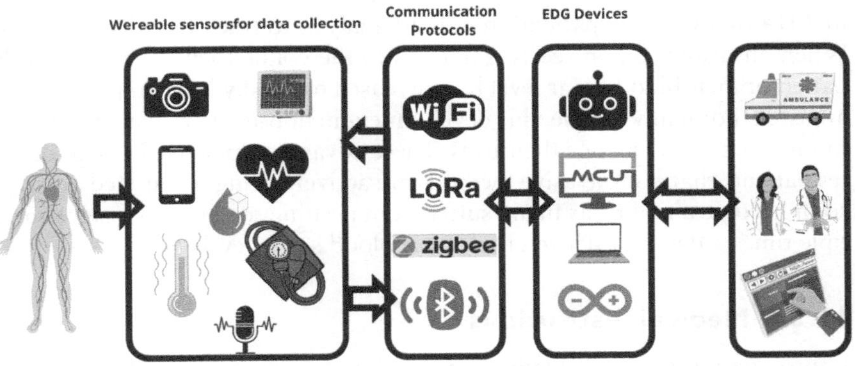

Figure 11.1 A general pipeline of a health monitoring system based on wearable devices.

levels, and oxygen saturation—to healthcare providers for evaluation. This enables clinicians to adjust treatment plans promptly without the need for physical visits. For example, patients recovering from surgery can be monitored at home via IoT devices that continuously track their vital signs. Alerts are automatically triggered in case of complications, allowing for immediate medical attention and reducing post-operative risks.

Furthermore, telemedicine has facilitated virtual consultations, allowing doctors and specialists to connect with patients through secure video platforms. These consultations are particularly beneficial for managing chronic conditions, where patients can receive regular care without traveling long distances. IoT devices integrated into telemedicine platforms not only enhance diagnostic precision but also enable real-time adjustments to treatment protocols based on continuous data flow. By reducing the reliance on in-person visits, telemedicine significantly lowers healthcare costs while maintaining high-quality care [3].

The integration of RPM and telemedicine exemplifies the transformative potential of IoT in healthcare. Together, these technologies enable personalized, flexible, and efficient patient management, fostering a healthcare model that is both proactive and responsive to individual needs.

11.2.3 Medical equipment

For customers of all ages, wearable medical gadgets are the most alluring alternative available for tracking their own vital signs in real time. There are currently more wearable gadgets available than Fitbit, Apple Watch, and the like. They record data, but they also carry out specific tasks in response to directions or recognized circumstances. "Intelligent associations" is one such. They have sensors built in to measure the extent of the underlying wound and identify whether an infection is present, whether it is healing, and whether topical medication is necessary [4].

In terms of health technology, "networked contact lenses" are another type of wearable. Google and Novartis started working on a linked contact lens in 2014 that used the patient's tear fluid analysis to track blood sugar levels. When an insulin pump receives data from the contact lenses, it notifies the patient if their blood sugar level has increased to a risky level and needs to be adjusted. For many people, this development in non-invasive diabetes patient monitoring could change their lives. These advancements give hope to diabetes patients that non-invasive methods are actively being researched and may soon become a reality, as many suffer from needing to prick themselves multiple times a day in order to check their blood sugar levels [12].

11.2.4 Medical institutions

Improving patient care quality is at the heart of many IoT advantages for the healthcare sector. But medical facilities have also improved as a result of the Internet of Things, for instance, by streamlining procedures and saving money. For instance, intelligent technology in healthcare institutions like hospitals makes sure that medical professionals can keep a closer eye on the effectiveness and lifespan of expensive equipment such as MRIs, CT and PT scanners, and X-ray machines. Mistakes or improper operation can be prevented in this way. The number of manual tests is reduced or sometimes completely eliminated with the use of remote sensors [13].

Time can now be allocated to more pressing duties. Relocating frequently utilized systems or equipment is a prevalent issue in medical facilities. If a device is lost in an emergency, this becomes a concern.

The integration of Bluetooth Low Energy (BLE) technology enables real-time device location tracking, ensuring rapid accessibility to critical tools during emergencies. This advancement not only alleviates stress in urgent situations but also enhances operational efficiency. Furthermore, even minor technological innovations have the potential to save numerous lives while incurring minimal costs. Among the various sectors influenced by the Internet of Things (IoT), healthcare has emerged as a domain where its transformative impact is particularly evident.

11.3 AI APPLICATIONS IN SMART HEALTHCARE

Healthcare has employed AI systems and AI-powered technologies for a range of tasks. The main requirement or objective in healthcare that AI may fulfill is defined by its purpose. The completion of a task highlights a specific function or process that the AI system facilitates or automates, emphasizing its role in enhancing efficiency and reducing manual effort. The prevalent forms of AI applications in healthcare are highlighted in a study of the literature. Here, we've categorized these apps according to how comparable their tasks and goals are. To illustrate the various applications of AI in smart healthcare, Figure 11.2 highlights the various tasks and objectives that AI can accomplish in this field.

- **AI for diagnostics**: To identify symptoms, patterns, illnesses, anomalies, and dangers, AI algorithms can evaluate data from wearable devices, electronic health records, and medical imaging (such as X-rays, MRIs, and CT scans). For example, AI can diagnose skin problems based on pictures of skin lesions, find diabetic retinopathy in eye scans, and detect malignancies in radiology scans [14, 15].
- **AI for personalized medicine**: By applying precision medicine techniques, AI can be used to provide more individualized and focused care by analyzing a patient's genetic information, lifestyle, medical history, and other unique factors. This allows for the provision of personalized treatment options, care plans, possible diagnoses, and drug dosage recommendations. For instance, AI is used by tech businesses

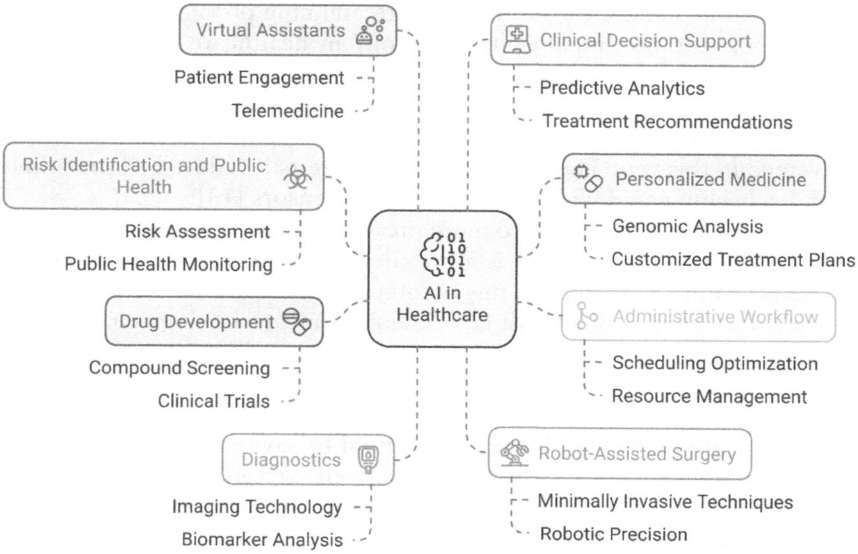

Figure 11.2 AI Applications in smart healthcare.

in this sector to tailor cancer therapy through precision medicine. They assist physicians in making data-driven decisions in real time by analyzing clinical and molecular data using machine learning and sophisticated bioinformatics [4, 16].

- **AI as virtual assistants**: AI-powered chatbots and voice assistants can be programmed to organize appointments, comprehend queries from patients, offer health advice, and function as an automated receptionist. Chatbots that assess symptoms, for example, can be used to treat chronic health conditions. Conversational chatbots, on the other hand, can operate as relational agents for mental health by offering emotional and mental health support. Additionally, wearable technology driven by AI and additional sensors are employed to track patients' health in real time.
- **AI as clinical decision support** (CDS): By assisting doctors in making better diagnoses and treatment decisions and based on risk assessments, guidelines, previous cases, and learning health patterns, AI systems might enhance doctors' understanding. Technology businesses in this sector, for instance, might focus on radiology AI by offering sophisticated imaging CDS tools that instantly identify acute irregularities in medical images. This helps radiologists rank cases according to urgency, which could hasten the identification and treatment of serious illnesses [17].
- **AI for drug development**: By evaluating molecular data, finding new medications, and streamlining clinical trials, AI can be used to speed

up and improve the efficiency of drug discovery. To speed up the usually expensive and time-consuming process of finding new medications, large databases, computing power, and predictive modelling are required. Technology businesses in this arena, for example, might concentrate on medicine discovery and aging research utilizing AI. Their artificial intelligence platform examines disease pathology and aging biology data to find novel targets for treatment. Deep learning is also utilized in the design of novel compounds for drug development and in the prediction of new medications' therapeutic uses [18].

- **AI for administrative workflow**: Automation of hospital administrative processes, including patient flow, bed availability, appointment scheduling, invoicing, and insurance authorization, can streamline efficiency and free up staff. By improving overall efficiency, this use of AI frees up healthcare staff to concentrate more on patient care and less on administrative duties. For instance, businesses that specialize in speech recognition technology can offer AI-driven solutions that facilitate the simplification of administrative duties in the healthcare industry. Their offerings include AI-driven coding and billing solutions that enhance the efficiency and precision of these procedures, as well as clinical documentation powered by AI that enables healthcare providers to promptly and properly record patient encounters [19].
- **AI for risk identification and public health**: By evaluating patient data, AI can be used to identify people who are at risk of contracting specific diseases, hence promoting preventive care. For instance, by examining trends in past patient data, such as vital signs, medical histories, and lifestyle variables, predictive models in healthcare AI can be used to identify patients at high risk of heart failure. By anticipating future outbreaks or the development of infectious diseases, AI can also be used to anticipate outbreaks [20].
- **Artificial intelligence as robot-assisted surgery and rehabilitation robots**: AI-guided robotic surgical devices can perform minimally invasive surgery more precisely than human surgeons alone, enabling complex surgeries. Furthermore, robots are made to help patients with physical therapy (such aiding those who have trouble walking) [21].

11.4 BENEFITS OF ARTIFICIAL INTELLIGENCE AND IoT IN SMART HEALTHCARE

The integration of Artificial Intelligence (AI) and the Internet of Things (IoT) in smart healthcare has introduced transformative advancements. These technologies synergistically enhance healthcare delivery by improving diagnostic accuracy, patient monitoring, operational efficiency, and cost management. This section outlines their key benefits [6, 22, 23].

- **Enhanced diagnostic accuracy**: The integration of IoT's data collection capabilities with AI's analytical power has revolutionized medical diagnostics. IoT devices, such as wearable health trackers and smart medical sensors, enable continuous and real-time collection of vital health data, including heart rate, blood pressure, and glucose levels. AI algorithms process this extensive data, identifying patterns and anomalies that assist in the early detection of diseases and prediction of health risks. For instance, IoT-enabled diagnostic systems combined with AI have demonstrated remarkable success in detecting chronic conditions such as diabetes and cardiovascular diseases. This combination accelerates diagnostic processes, reduces human errors, and improves patient outcomes by enabling timely and precise interventions.
- **Proactive monitoring and preventive care**: IoT devices facilitate continuous patient monitoring, allowing for the collection of real-time physiological data, while AI systems analyze this data to anticipate potential health complications. This proactive approach shifts healthcare from a reactive to a preventive model, reducing the incidence of critical medical events.

  For example, IoT-connected cardiac monitors and respiratory sensors continuously track patient vitals. AI-driven analytics interpret this information, enabling healthcare providers to intervene before a situation escalates. Such integration has proven especially beneficial for managing chronic diseases, minimizing hospital readmissions, and enhancing the quality of life for elderly and vulnerable populations.
- **Optimization of hospital processes**: AI and IoT technologies streamline hospital operations by optimizing resource management and reducing inefficiencies. IoT devices track medical equipment, manage bed availability, and monitor energy consumption, while AI automates administrative tasks like appointment scheduling and staff allocation.

  During periods of high demand, such as pandemics [24], AI and IoT systems predict resource needs and optimize workflows, ensuring that medical staff can focus on patient care rather than logistical challenges. This integration enhances overall operational efficiency and improves the patient experience.
- **Cost reduction and efficiency improvement**: The combined implementation of AI and IoT reduces healthcare costs by minimizing inefficiencies and enabling more effective treatments. IoT-based telemedicine platforms, for instance, decrease the need for in-person consultations, while AI optimizes treatment plans based on patient-specific data, reducing the likelihood of ineffective therapies.

  Additionally, continuous monitoring via IoT devices helps prevent costly complications by detecting health issues early. These savings extend to healthcare providers, who can allocate resources more effectively and lower operational costs without compromising the quality of care.

- **Advancements in personalized medicine**: IoT and AI are central to the development of precision medicine, where treatments are customized based on individual patient data. IoT devices collect real-time insights about a patient's physiology, which AI then analyzes to provide tailored therapeutic recommendations. For example, IoT-enabled wearables track patient responses to medication, and AI algorithms adjust drug dosages dynamically, minimizing adverse effects. This personalized approach improves treatment outcomes and fosters patient-centered care.

The synergy between Artificial Intelligence and the Internet of Things is transforming healthcare systems globally. By integrating real-time data collection with advanced analytical tools, these technologies enable a proactive, personalized, and sustainable healthcare model. Their combined impact enhances diagnostic accuracy, operational efficiency, and cost management, while paving the way for innovative approaches to patient care. As the integration of AI and IoT continues to evolve, it holds the potential to redefine the standards of modern healthcare and address the most pressing challenges in the field.

11.5 CHALLENGES OF AI AND IoT IN SMART HEALTHCARE

The integration of Artificial Intelligence (AI) and the Internet of Things (IoT) into smart healthcare offers transformative opportunities but also presents significant challenges that must be addressed to fully unlock their potential. This section outlines the key challenges identified in recent studies.

- **Data protection and privacy**: AI and IoT rely on extensive data collection and processing, raising critical concerns about data security and patient privacy. IoT devices continuously collect sensitive health information, such as heart rate, glucose levels, and oxygen saturation, while AI analyzes this data to generate actionable insights. Ensuring secure transmission, storage, and usage of this data is essential to prevent breaches and maintain patient trust. For instance, IoT devices are susceptible to cyberattacks, potentially compromising patient data or disrupting critical healthcare operations. Similarly, AI systems require robust data governance frameworks to ensure compliance with privacy regulations and to address risks related to unauthorized data access [11, 25].
- **Ethical questions and accountability**: The adoption of AI and IoT technologies in healthcare raises ethical concerns, including informed consent, transparency, and accountability for automated decisions. IoT systems must ensure patients are fully aware of how their data

is collected, used, and shared. On the AI side, biases in algorithms can lead to discriminatory outcomes, further complicating the ethical landscape. Moreover, the accountability for decisions made by interconnected systems, such as IoT devices feeding data to AI algorithms, remains unclear. Establishing clear legal and regulatory frameworks is essential to address these issues and build trust among stakeholders [5].

- **Integration and adoption**: Healthcare professionals often struggle to adopt AI and IoT technologies due to limited familiarity or trust in these innovations. Practitioners may hesitate to rely on automated systems for critical decisions, especially when these systems integrate IoT devices and AI algorithms. Addressing this challenge requires comprehensive training programs, user-friendly interfaces, and collaborative efforts between technologists and healthcare providers to facilitate acceptance and effective utilization [26].
- **Development and regulation**: The development and deployment of AI and IoT systems must comply with stringent regulatory standards to ensure safety and efficacy. For IoT devices, certification processes for hardware reliability and cybersecurity are critical. For AI systems, clinical validation and algorithmic transparency are mandatory prior to implementation. The absence of unified international regulations exacerbates these challenges, as different regions impose varying compliance requirements, complicating the global deployment of these technologies.
- **Technical challenges**: The seamless integration of AI and IoT in healthcare systems faces numerous technical barriers.

For IoT: Managing interoperability among diverse devices, ensuring reliable connectivity in remote areas, and addressing device failures or malfunctions.

For AI: Training models on heterogeneous and complex medical datasets, ensuring algorithm robustness, and maintaining adaptability to rapidly evolving medical environments.

The combination of IoT and AI amplifies these challenges, as the quality and consistency of IoT-generated data directly affect the accuracy and reliability of AI predictions.

11.6 SECURITY ATTACKS IN SMART HEALTHCARE SYSTEMS

In the realm of smart healthcare systems, security attacks pose significant threats that can compromise patient care and data integrity. As outlined in Table 11.1, various types of attacks include Denial of Service, which floods networks with excessive traffic, rendering critical healthcare services

Table 11.1 Classification of cybersecurity attacks and their impact on smart healthcare systems

Art	*Attack type*	*Description*	*Example consequences*
[27]	Denial of service	Attack aimed at rendering critical services inaccessible by flooding the network with packets.	Inaccessibility to medical data, delays in healthcare delivery.
[13]	Data breach	Unauthorized access to sensitive information such as patients' medical records.	Disclosure of personal data, risk of identity theft.
[28]	Phishing	Attempt to steal sensitive information by creating fraudulent web links with malicious code.	Theft of identification data, compromise of accounts.
[6]	Malware	Attacks aimed at compromising electronic patient record systems	Data corruption, interruption of healthcare services.
[29]	Man-in-the-Middle	Interception of communication between legitimate users to steal information.	Theft of medical data, alteration of critical data.
[30]	Software-based	Exploitation of software vulnerabilities, including outdated and unpatched software.	Exposure to security flaws, risk of data breach.
[31]	Side-channel	Attacks aimed at extracting sensitive information by exploiting side-channel information.	Extraction of sensitive data, compromise of confidentiality.

inaccessible and leading to delays in medical data availability and healthcare delivery. Data breaches allow unauthorized access to sensitive patient records, elevating the risk of identity theft and personal data disclosure. Phishing attacks deceive individuals into disclosing confidential information, thereby compromising user accounts and patient identification data. Malware infections specifically target electronic patient record systems, resulting in data corruption and interruptions to healthcare services. Man-in-the-Middle attacks intercept communications between users to steal or alter critical medical information. Software-based vulnerabilities exploit flaws in outdated or unpatched software, increasing susceptibility to breaches. Finally, side-channel attacks extract sensitive information by exploiting indirect data paths within the system, such as electromagnetic emissions or power analysis, thereby breaching the confidentiality of critical healthcare data.

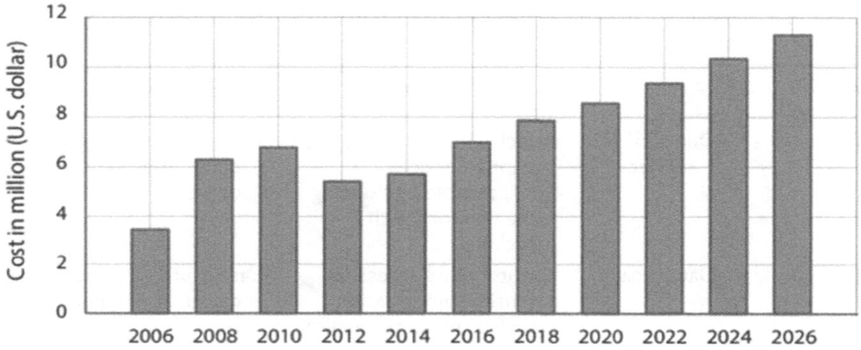

Figure 11.3 Total financial loss due to cyberattacks on healthcare industry (2006–2026) [32].

Figure 11.3, depicted alongside Table 11.1 as a diagram, illustrates the financial loss due to cyberattacks on the healthcare industry from 2006 to 2026. Data breaches have the most significant impact on the industry, with losses amounting to 8.7 million in the current year alone, harming the national economy. The diagram's forecasted data emphasize the imperative for robust security measures, such as integrating various technologies and safeguards, to detect and mitigate potential threats before they jeopardize the continuity and safety of patient care in smart healthcare systems.

11.7 DISCUSSION

The integration of artificial intelligence (AI) and the Internet of Things (IoT) has undoubtedly transformed the healthcare landscape, offering unprecedented opportunities for advancements in medical diagnosis, treatment, and patient care. The potential benefits of smart healthcare are vast, ranging from personalized treatment plans to predictive analytics that can enhance patient outcomes [6].

However, the adoption of AI and IoT in healthcare also brings forth significant challenges, particularly in terms of data privacy and security. Addressing these concerns is critical to ensuring the trust and safety of patients while utilizing these transformative technologies. As researchers and developers continue to innovate, collaboration between stakeholders such as healthcare providers, policymakers, and technology experts will be essential to establish robust standards and best practices.

Looking ahead, the future of smart healthcare appears promising. Advances in machine learning and data analytics will likely drive further innovation, enabling more precise and effective medical care. Additionally, the ongoing development of secure and scalable solutions will help mitigate risks associated with the implementation of AI and IoT in healthcare.

As the field continues to evolve, it will be crucial to balance the potential benefits of AI in smart healthcare with ethical considerations and patient-centric approaches. By fostering a collaborative and transparent environment, the medical industry can harness the full potential of AI and IoT to improve health outcomes and revolutionize patient care.

11.8 CONCLUSION

The integration of artificial intelligence (AI) and the Internet of Things (IoT) into healthcare signifies a paradigm shift with far-reaching implications for patient care and clinical practices. These technologies enable precision medicine, enhance diagnostic accuracy, and facilitate continuous health monitoring, leading to improved patient outcomes and operational efficiencies. However, this digital transformation is accompanied by substantial challenges, particularly concerning data security, privacy, and ethical governance.

Addressing these issues requires a multifaceted approach that includes the development of robust cybersecurity frameworks, regulatory policies, and interdisciplinary collaborations among technologists, clinicians, and policymakers. The successful adoption of AI and IoT in healthcare demands not only technological innovation but also a steadfast commitment to ethical practices that prioritize patient rights and trust.

As healthcare systems evolve, the harmonious integration of innovation with ethical and legal safeguards will be instrumental in ensuring sustainable progress. Embracing AI and IoT responsibly can unlock unprecedented opportunities for smart healthcare, setting the stage for a future where technology and medicine converge to prioritize human well-being and societal health.

REFERENCES

[1] Healthcare IoT - Worldwide | Statista Market Forecast, Statista. Consulté le: 21 décembre 2024. [En ligne]. Disponible sur: https://www.statista.com/outlook/tmo/internet-of-things/healthcare-iot/worldwide

[2] Gong, F. F., Sun, X. Z., Lin, J., & Gu, X. D. (2013), Primary exploration in establishment of China's intelligent medical treatment. *Modern Hospital Management* 11(2), 28–29.

[3] Shafi, J., et al. Waheed, Role of Smart Wearable in Healthcare: Wearable Internet of Medical Things (WIoMT), in *The IoT and the Next Revolutions Automating the World*, IGI Global Scientific Publishing, 2019, p. 133–155. doi: 10.4018/978-1-5225-9246-4.ch009

[4] Kelley, S. O., Challenges and Opportunities for Wearable Sensing Systems, *ACS Sens.*, vol. 7, no 2, p. 345–346, févr. 2022, doi: 10.1021/acssensors.2c00284
[5] Memarzadeh, A. B. Et K. Chapter 2 - The rise of artificial intelligence in healthcare applications, in *Artificial Intelligence in Healthcare*, A. Bohr Et K. Memarzadeh, Éd., Academic Press, 2020, p. 25–60. doi: 10.1016/B978-0-12-818438-7.00002-2
[6] Smart Healthcare in the Age of AI: Recent Advances, Challenges, and Future Prospects | *IEEE Journals & Magazine | IEEE Xplore. Consulté le*: 5 mai 2024. [En ligne]. Disponible sur: https://ieeexplore.ieee.org/abstract/document/9565155
[7] Mutlag, A. A., et al., MAFC: Multi-Agent Fog Computing Model for Healthcare Critical Tasks Management, *Sensors*, vol. 20, no 7, Art. no 7, janv. 2020, doi: 10.3390/s20071853
[8] Miao, F., Liu, Z.-D., Liu, J.-K., Wen, B., He, Q.-Y., & Li, Y., Multi-Sensor Fusion Approach for Cuff-Less Blood Pressure Measurement, *IEEE Journal of Biomedical and Health Informatics*, vol. 24, no 1, p. 79–91, janv. 2020, doi: 10.1109/JBHI.2019.2901724
[9] Yang, F., Zhao, X., Jiang, W., Gao, P., & Liu, G., Multi-method Fusion of Cross-Subject Emotion Recognition Based on High-Dimensional EEG Features, *Front. Comput. Neurosci.*, vol. 13, août 2019, doi: 10.3389/fncom.2019.00053
[10] Gu, Q., Jiang, S., Lian, M., & Lu, C., Health and Safety Situation Awareness Model and Emergency Management Based on Multi-Sensor Signal Fusion, *IEEE Access*, vol. 7, p. 958–968, 2019, doi: 10.1109/ACCESS.2018.2886061
[11] Realizing an Effective COVID-19 Diagnosis System Based on Machine Learning and IoT in Smart Hospital Environment | *IEEE Journals & Magazine | IEEE Xplore. Consulté le*: 5 mai 2024. [En ligne]. Disponible sur: https://ieeexplore.ieee.org/abstract/document/9319693
[12] Bloch-Budzier, S. (2016). *NHS using Google technology to treat patients*. BBC News, 22.
[13] How Security Vulnerabilities Pose Risks for Healthcare Organizations. Available online: https://www.techrepublic.com/article/ se-curity-vulnerabilities-healthcare/ (accessed on 21 August 2019).
[14] Castiglioni, I. et al., AI applications to medical images: From machine learning to deep learning, *Physica Medica*, vol. 83, p. 9–24, mars 2021, doi: 10.1016/j.ejmp.2021.02.006
[15] Tremblay, P. H. Et J., Artificial intelligence in medicine, *Metabolism*, vol. 69, p. S36–S40, avr. 2017, doi: 10.1016/j.metabol.2017.01.011
[16] Precision Medicine, AI, and the Future of Personalized Health Care - Johnson - 2021 - Clinical and Translational Science - Wiley Online Library. Consulté le: 5 mai 2024. [En ligne]. Disponible sur: https://ascpt.onlinelibrary.wiley.com/doi/full/10.1111/cts.12884
[17] Bizzo, B. C., Almeida, R. R., Michalski, M. H., & Alkasab, T. K., Artificial Intelligence and Clinical Decision Support for Radiologists and Referring Providers, *Journal of the American College of Radiology*, vol. 16, no 9, Part B, p. 1351–1356, sept. 2019, doi: 10.1016/j.jacr.2019.06.010
[18] Molecular representations in AI-driven drug discovery: a review and practical guide | Journal of Cheminformatics. Consulté le: 5 mai 2024. [En ligne]. Disponible sur: https://link.springer.com/article/10.1186/s13321-020-00460-5

[19] The potential for artificial intelligence in healthcare - PMC. Consulté le: 5 mai 2024. [En ligne]. Disponible sur: https://www.ncbi.nlm.nih.gov/pmc/articles/PMC6616181/

[20] Applications of artificial intelligence and machine learning in heart failure | European Heart Journal - Digital Health | Oxford Academic. Consulté le: 5 mai 2024. [En ligne]. Disponible sur: https://academic.oup.com/ehjdh/article/3/2/311/6585514

[21] Sqalli, M. T. & Al-Thani, D., AI-supported Health Coaching Model for Patients with Chronic Diseases, in *2019 16th International Symposium on Wireless Communication Systems (ISWCS)*, août 2019, p. 452–456. doi: 10.1109/ISWCS.2019.8877113

[22] IOT and AI in Healthcare: A Systematic Literature Review, *IIS*, 2018, doi: 10.48009/3_iis_2018_33-41

[23] Gupta, R., Patel, M. M., Tanwar, S., Kumar, N., & Zeadally, S., Blockchain-Based Data Dissemination Scheme for 5G-Enabled Softwarized UAV Networks, *IEEE Transactions on Green Communications and Networking*, vol. 5, no 4, p. 1712–1721, déc. 2021, doi: 10.1109/TGCN.2021.3111529

[24] Aissaoui, W., Khennou, F., & Abdellaoui, A. (2023, December). Enhancing Intensive Care Patient Prognostics with Machine Learning. In *Proceedings of the 12th International Symposium on Information and Communication Technology* (pp. 546–553).

[25] Rahman, S. M. A., Ibtisum, S., Podder, P., & Hossain, S. M. S., Progression and Challenges of IoT in Healthcare: A Short Review, *IJCA*, vol. 185, no 37, p. 9–15, oct. 2023, doi: 10.5120/ijca2023923168

[26] Besenyő, J. & Kovács, A. M., Healthcare cybersecurity threat context and mitigation opportunities, *Security Science Journal*, vol. 4, no 1, p. 83–101, Art. no 1, avr. 2023.

[27] HC3 Warns Healthcare Sector of Karakurt Ransomware Group. Available online: https://healthitsecurity.com/news/hc3-warnshealthcare-sector-of-karakurt-ransomware-group (accessed on 12 July 2022).

[28] Phishing Attacks, Email Security Incidents Hit 3 Healthcare Orgs. Available online: https://healthitsecurity.com/news/phishingattacks-email-security-incidents-hit-3-healthcare-org (accessed on 14 July 2022).

[29] Salem, O., Alsubhi, K., Shaafi, A., Gheryani, M., Mehaoua, A., & Boutaba, R. (2021). Man-in-the-Middle attack mitigation in internet of medical things. *IEEE Transactions on Industrial Informatics*, *18*(3), 2053–2062.

[30] Mejía-Granda, C. M., Fernández-Alemán, J. L., Carrillo-de-Gea, J. M., & García-Berná, J. A. (2024). Security vulnerabilities in healthcare: an analysis of medical devices and software. *Medical & Biological Engineering & Computing*, *62*(1), 257–273.

[31] Man, K., Wang, Z., Hao, Y., Zheng, S., Zhou, X. A., Cao, Y., & Qian, Z. (2024, November). SCAD: Towards a Universal and Automated Network Side-Channel Vulnerability Detection. In *2025 IEEE Symposium on Security and Privacy (SP)* (pp. 68–68). IEEE Computer Society.

[32] Cost of a Data Breach 2022. Available online: https://www.ibm.com/reports/data-breach (accessed on 12 September 2022).

Chapter 12

Fortifying Industrial IoT (IIoT)

Leveraging AI for optimizing security

Shahad AL-Tamimi and Qasem Abu Al-Haija

12.1 INTRODUCTION

Nowadays Industrial Internet of Things (IIoT) has drastically altered industrial processes through the possibilities for networked systems, smooth data sharing, and increased automation. IIoT expands the capabilities of the IoT to industrial applications, revolutionizing sectors that include manufacturing, energy, and healthcare. This change has increased productivity and creativity while also posing substantial cybersecurity issues owing to the variety of devices, outdated systems, and the enormous attack surface produced by networked settings.

For the importance of industrial processes, securing the security of IIoT systems is crucial. Cyberattacks on IIoT systems, such as ransomware, advanced persistent threats, and data breaches, may cause catastrophic interruptions. This has generated an urgent demand for strong security measures. When IIoT integrates with AI, its ability to analyze massive volumes of data and react to attacks in real time has emerged as a critical tool for improving IIoT security. IIoT systems may become more resilient and adaptable to changing cyber threats by incorporating AI-driven technologies such as ML, blockchain, and edge computing.

12.1.1 IoT and IIoT

IoT, IIoT, Cyber-Physical Systems (CPS), and Industry 4.0 are interrelated ideas driving current technology. Integrating physical systems with computational intelligence, CPS enables real-time monitoring and control. Where the key relation between IoT, IIoT, Industry 4.0 and industry internet is illustrated in Figure 12.1. IoT expands device connection to enable data interchange across applications. IIoT increases industrial output, energy, and logistics via smart, networked systems. Moreover, IIoT enhances industrial operations via big data, analytics, and ML. Table 12.1 shows the differences between IoT and IIoT. The term IoT is connected to CPS, Industry 4.0, and the industrial internet. Helen Gill's 2006 CPS idea merges sensing and embedded systems, merging software and hardware, to enable efficient

 DOI: 10.1201/9781003606307-12

Figure 12.1 The relationships between CPS, IoT, IIoT, industrial internet, and Industry 4.0.

Table 12.1 Examination of the major features of IoT vs IIoT

Characteristics	*IoT*	*IIoT*
Framework	Self-reliant	Industrial facility-reliant
Applications	Intelligent home, health tracking, and localization of interiors.	Smart solutions for logistics, manufacturing, distant preservation, and transportation.
Size of Development	Small	Large
Mobility	High	Low
Data volume	Medium	High
Delay sensitivity	High	Low

internal information flow, real-time feedback, and positive communication between virtual and physical things [1]. IoT is a subset of CPS which enables the communication between various things over the internet using unique IDs. The internet supports IoT devices by enabling availability, interoperability, universality, and socializing. Germany launched Industry 4.0. The global concept of CPS and emerging technologies, such as AI, IoT and big data, are used to create intelligent manufacturers.

Hussain et al. [2] highlight the importance of the IIoT in promoting Industry 4.0, which encompasses smart manufacturing and industrial automation. However, it also emphasizes the rising potential of cyberattacks, including Advanced Persistent Threats (APTs) and botnets, which may damage IIoT networks. Moreover, the researchers of the study present a deep learning-enabled hybrid framework for effectively detecting and mitigating these risks, displaying excellent detection accuracy with no impact on performance.

In summary, CPS connects the physical and digital worlds, while IoT enables the communication between physical devices in both civilian and industrial settings. Also, IIoT uses the development technology to forecast and respond to future trends. As subsets of CPS, IoT, and IIoT work together to power industrial applications, with IIoT concentrating primarily on improving industrial processes. In addition, Industry 4.0 combines IIoT with other cutting-edge technologies to develop smart, efficient, and flexible industrial processes, thus anticipating a highly automated, intelligent, and digitally connected IIoT that will transform production and service delivery. These interconnected concepts serve as the foundation for the present digital revolution in a range of fields.

12.1.2 Motivation

The main aspect of this chapter is motivated by the critical need to protect IIoT systems in the face of escalating cyber threats. It contributes by evaluating current advances in AI-enabled security, with a focus on deep learning, federated learning, and blockchain integration. Furthermore, it shows how adaptive and decentralized security solutions may enhance IIoT resilience. This chapter aims to pave the way for safe IIoT adoption by tackling new security challenges while ensuring data integrity, confidentiality, and availability in industrial contexts.

12.1.3 Contribution

This chapter contributes to the growing body of research on IIoT security by offering:

I. Examination of the existing cybersecurity issues related with IIoT, including device vulnerabilities, protocol flaws, and advanced attacks.
II. Examination of AI's role in improving IIoT security via better threat detection, adaptive defensive mechanisms, and integration with blockchain and edge computing technologies within 2.1 and 2.2.
III. Overview of IIoT security trends and problems, with practical insights for academics and practitioners working to create safe and efficient industrial ecosystems.

By tackling these issues, this chapter hopes to plug the gap among theoretical advances in AI-driven security and their actual use in IIoT systems.

12.1.4 Chapter organization

Additionally, this chapter includes a discussion on the principles and fundamentals of Iot and IIoT. The chapter also explores the motivation behind this and provides a summary of its major contributions and motivation within Section 12.1. The subsequent sections of this chapter are structured as follows. Section 12.2 presents the latest trends within IIoT organized into main categories: edge computing security, AI-driven and ML for cybersecurity, deep learning, blockchain technology, privacy-preserving and federated learning, and lastly, 5G-enhanced IIoT security. Section 12.3 provides the Industrial Internet of Things (IIoT) via cybersecurity. Thus, Section 12.4 contains open challenges and potential solutions for IIoT. Section 12.5 provides details on legal regulatory governing security and privacy for IIoT. Finally, Section 12.6 concludes the chapter.

In short, the chapter provides a substantial and well-documented background on IIoT, integrating a historical perspective with a review of latest trends, followed by the integration of IIoT via AI environment and cybersecurity.

12.2 TRENDS OF SECURE IIoT

IIoT has transformed industrial processes with networked devices, real-time data transmission, and improved automation. IIoT systems boost efficiency and innovation in manufacturing, healthcare, and energy. IIoT integration increases cybersecurity risks since connection increases the attack surface. IIoT security evolves to meet various challenges. The latest trends for IIoT are shown in Figure 12.2. Thus, this section illustrates the IIoT security trends to reduce risks, improve dependability, and meet the complexity of a hyper-connected industrial environment. Also, understanding these patterns allows firms to deploy proactive security measures to safeguard IIoT infrastructure from existing and upcoming threats.

12.2.1 Edge computing security

Alotaibi [3] discusses the significance of edge computing security within the context of the IIoT. As data processing moves closer to devices, protecting edge computing environments is growing critical. This involves implementing lightweight encryption and local threat detection systems. Thus, the study emphasizes the necessity for adequate protection frameworks to guard against possible attacks within edge devices and applications. Highlighting AI-based solutions and edge security may help to reduce risks, resulting in safer and more robust IIoT environments. Furthermore, the research by Sasikumar et al. [4] highlights that IIoT and edge computing interact to provide a safe and energy-efficient consensus mechanism. This technique uses

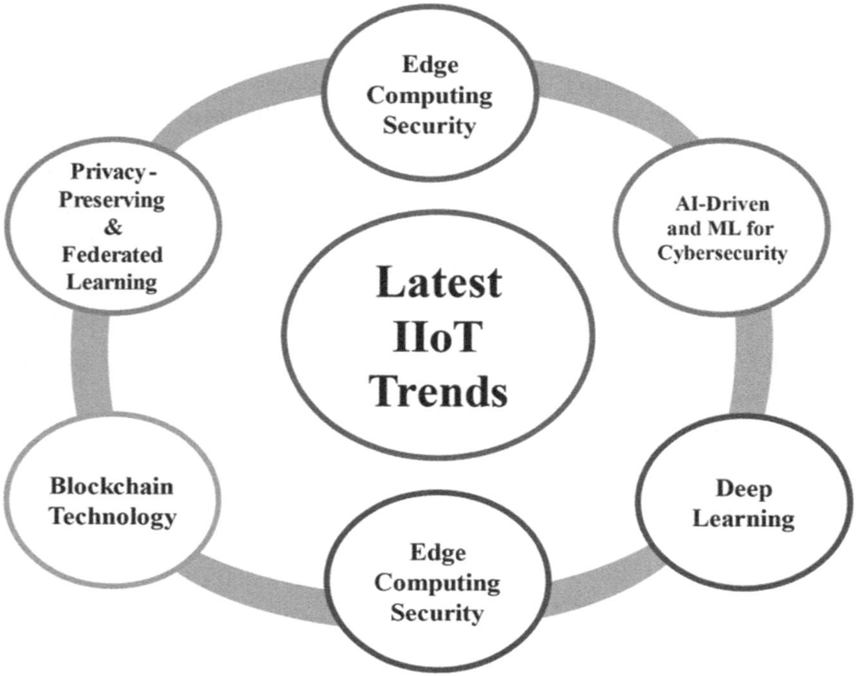

Figure 12.2 IIoT latest trends.

AI to ensure the sustainability and efficiency of smart industrial settings. By processing data at the edge, closer to the source, the system minimizes latency, improves reaction times, and optimizes energy usage, enhancing overall security and operational efficiency in IIoT applications. Czeczot et al. [5] discuss how AI can manage cybersecurity for Industry 4.0 and Industry 5.0 using IIoT. Edge computing may improve threat detection and response by processing data locally, lowering latency, and increasing real-time decision-making. Finally, Jiang et al. [6] examine how AI-enabled SDN technologies might enhance industrial IoT network security and functionality. Edge computing is important for effective data processing and administration, providing industrial security and functioning.

12.2.2 AI-driven and ML for cybersecurity

Trakadas et al. [7] also discuss AI-based cooperation in industrial IoT production, including essential principles, architectural extensions, and prospective applications. Edge computing is essential for real-time data analysis and industrial system collaboration. In addition, the study by Lv et al. [8] discusses on AI-based industrial IoT system dependability. It describes how edge computing might improve system dependability and security by processing data locally and decreasing cloud dependence.

12.2.3 Deep learning

The study by Shahin et al. [9] explores AI-enabled Intrusion Detection Systems (IDS) for enhancing network security in the IIoT. It emphasizes the growing trend of using deep learning techniques to detect and mitigate sophisticated cyber threats in real-time, ensuring the protection of IIoT environments against advanced and frequent cyberattacks. In addition, the study by Yazdinejad et al. [10] offers an ensemble deep learning model toward IIoT cyber threat hunting which employs LSTM and AE architectures to identify abnormalities and increase accuracy. The study by Latif et al. [11] explores IIoT-related deep learning algorithms, their potential uses, implementation frameworks, and opportunities for the future.

These research efforts demonstrate the expanding use of deep learning to solve IIoT security and efficiency issues, demonstrating the potential for sophisticated AI approaches to enhance anomaly detection and system performance. Also, researchers stress deep learning's role in IIoT system efficiency and security.

12.2.4 Blockchain technology

To begin exploring blockchain technology within IIoT, the study [12] provides a lightweight blockchain security architecture to improve IIoT security and privacy, where the solution relies on blockchain's decentralization and immutability to protect data and prevent cyberattacks. Moreover, the study [13] also tackles IIoT privacy and the potential of blockchain technology in addressing it. Blockchain's openness and immutability ensure the integrity and confidentiality of IIoT data. In another study [14], the authors examine the application of blockchain-based AI methods in IIoT administration, including current advancements, integration issues, and future possibilities. AI and blockchain technologies enhance IIoT security, trust, and efficiency. Finally, AI-powered IIoT security and trust can be enhanced via blockchain, ensuring data integrity, transparency, and secure communication across industrial networks. The study by Zhang et al. [15] highlights how blockchain's decentralized and transparent nature fosters security and trust among networked devices.

12.2.5 Privacy-preserving and federated learning

For privacy-preserving, the study by Chen et al. [16] explores the integration of privacy-preserving and traceable federated learning for data sharing in industrial IoT applications. Also, the authors' focus is on ensuring data privacy and traceability, which are critical for secure and trustworthy

IIoT systems. Furthermore, the study by Arachchige et al. [17] proposes a trustworthy privacy-preserving framework for ML in IIoT systems. The framework attempts to improve data privacy and security while preserving the efficiency and efficacy of machine learning models. However, the study by Fu et al. [18] presents VFL (Verifiable Federated Learning), a paradigm intended to enable privacy-preserving and verifiable data processing for massive data in IIoT. The emphasis is on secure, decentralized data handling to protect sensitive information. The study by Nguyen et al. [19] discusses the role of federated learning in the future of industrial IoT, highlighting its potential to improve data privacy and security without sacrificing the performance of IIoT applications. Federated learning permits data to be processed locally, limiting the likelihood of a data leak. Ultimately, the study by Ruzafa-Alcázar et al. [20] focuses on intrusion detection in the IIoT using privacy-preserving federated learning. It emphasizes the necessity of protecting data privacy while successfully recognizing and mitigating cyber risks in industrial contexts.

12.2.6 5G-enhanced IIoT security

The research by Mukherjee et al. [21] investigates how big data analytics might improve the security of 5G-enabled IoT and IIoT systems, hence promoting the development of sustainable smart cities. The study intends to increase data processing and threat detection by making use of 5G networks' high speed and low latency, leading to more reliable and secure smart city infrastructures.

12.2.7 Other trends

The authors of the articles listed in the bibliography [22, 23] assert the use of Extreme Learning Machines (ELMs) to enhance the IDS for IoT and IIoT networks. ELMs have substantial advantages owing to their capability of accommodating high-dimensional data and enhancing accuracy of the detection; thus, they are a good approach toward real-time threat identification. The complementary feature that both studies emphasize is the ability of ELMs to quickly process and analyze huge volumes of information, which helps to ensure high-level defense against cyber threats in IIoT environments and better cope with their increasing complexity.

12.3 INDUSTRIAL INTERNET OF THINGS (IIoT) VIA CYBERSECURITY

IIoT permits networked devices, sensors, and systems collect, distribute, and analyze data in real time, enhancing efficiency and innovation. IIoT adoption increases cybersecurity risks. Legacy system integration, networked devices' large attack surface, and vital industrial processes make them cyber-attack targets. Operational integrity, data protection, and critical infrastructure disruption prevention need IoT security. AI, blockchain, and strong encryption standards can help organizations construct IIoT ecosystems that can survive cyberattacks and support innovation and trust in a digital-first future. To defend IIoT systems against cyberattacks, the authors highlight on cybersecurity. IIoT boosts industrial efficiency and innovation, and thus data privacy is crucial [24].

12.3.1 IIoT architecture

IIoT architecture is represented in Figure 12.3. IIoT integrates devices, networks, infrastructure, and applications to provide intelligent industrial solutions. Smart schools, hospitals, homes, and communities produce data using sensors and other IoT-enabled components at the base layer. Ethernet, Bluetooth, Wi-Fi, cellular networks, and NFC networks send this data for dynamic and flexible communication, backed by Software-Defined Networking (SDN) and Network Function Virtualization (NFV). Actionable insights rely on infrastructure layer data processing, detection, audio/video analysis, and device/data management. IIoT delivers monitoring, condition examination, security control, and quality control at the application layer in order to improve operational efficiency, security, and dependability. Layered design facilitates sophisticated application communication, management, and deployment in many industrial settings.

In general, IIoT architecture and security are changing. Alrawashdeh et al. [25] stress the significance of Industrial Identity Management Systems for IIoT security. The study by Pivoto et al. [26] evaluates cyber-physical system designs and their relevance in Industry 4.0 IoT integration. In Khowaja et al. [27], they offer a two-tier system for IIoT data and model security using federated learning and encryption. These studies demonstrate the necessity for strong security to secure linked industrial systems and guarantee dependable and efficient operations.

12.3.2 IIoT/ IoT via CIA triad

Contemporary companies are increasingly dependent on the IIoT and IoT, but their expansion presents various major security problems that must be addressed to maintain system integrity. Security for IIoT and IoT ecosystems starts with the Confidentiality, Integrity, and Availability (CIA) triad which is shown in Figure 12.4. IIoT systems that manage significant industrial data need confidentiality to secure sensitive data. Mobile and industrial IoT data confidentiality may be improved using biometric authentication and encryption [27]. IIoT systems that use real-time data for decision-making need integrity to assure data accuracy and unaltered transmission. Secure bot-resistant network topologies protect edge-enabled IIoT data [28]. IIoT

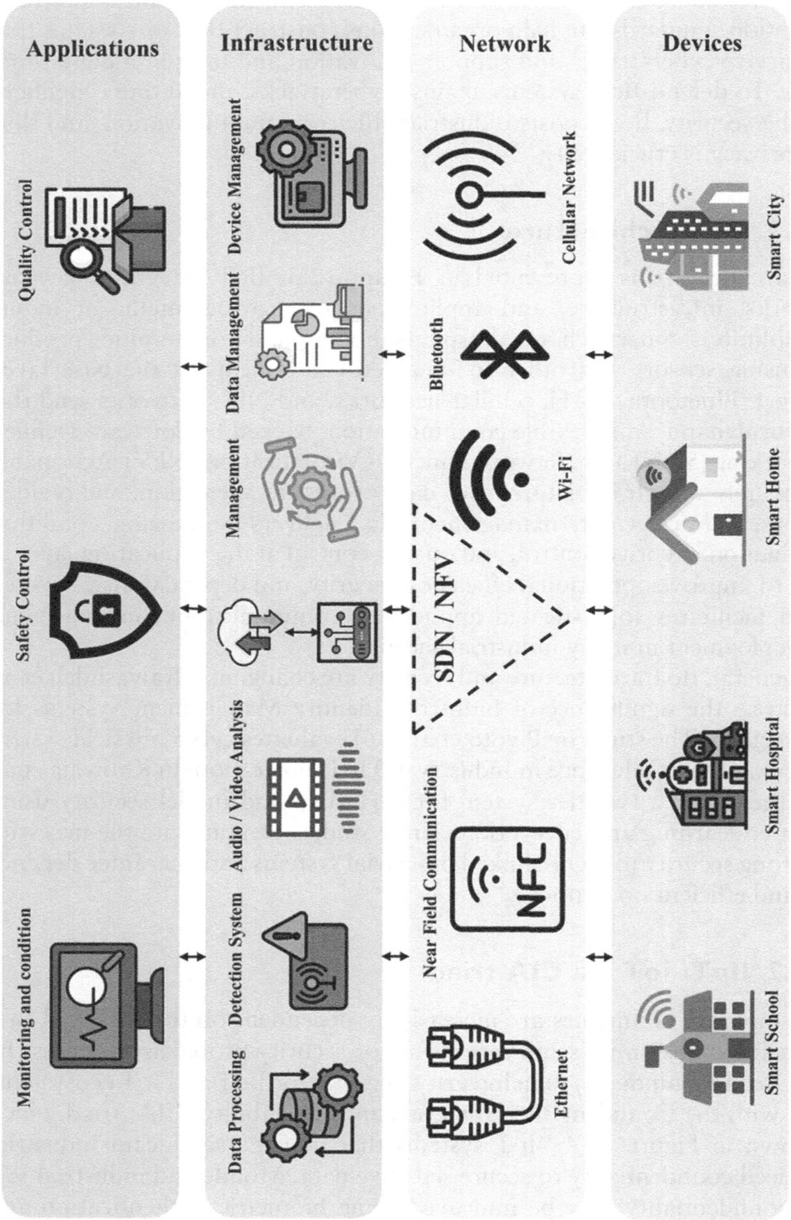

Figure 12.3 The system architecture of IIoT.

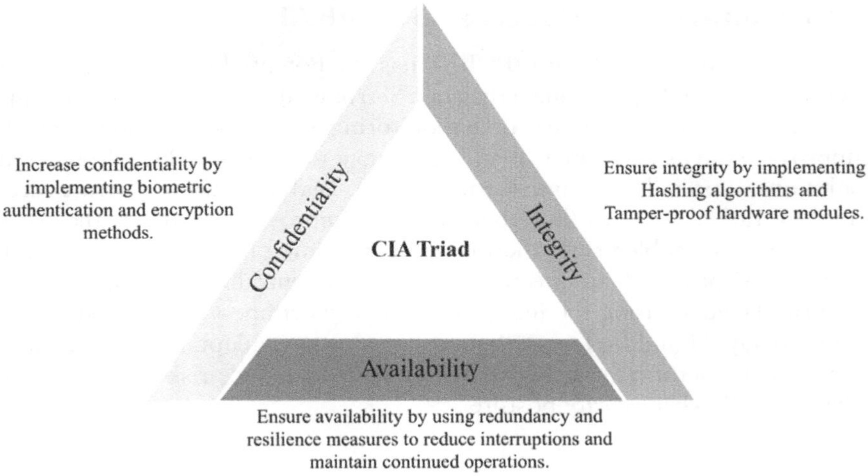

Figure 12.4 IIoT via CIA triad.

device operation and services is vital for industrial productivity and safety. Redundancy and sturdy architecture guard against cyberattacks and system breakdowns [29]. The CIA trinity principles for a secure and dependable operational architecture may help IIoT and IoT systems address escalating cybersecurity issues.

In short, protecting the confidentiality, integrity, and availability (CIA) triad of IIoT and IoT systems is essential for their usefulness and dependability in industrial and commercial settings. Khowaja et al. [27] note that fast IIoT technology improvements have created security vulnerabilities that need strong countermeasures to secure sensitive data and activities. Biometric authentication improves confidentiality and prevents unwanted access in mobile and industrial IoT ecosystems, according to Tan and Samsudin [28].

12.3.3 Cybersecurity risks

Advanced industrial systems leverage the IIoT to connect devices, machines, and networks for efficiency, innovation, and automation. Interconnected IIoT systems pose serious cybersecurity risks to essential functions and data confidentiality, integrity, and availability. Outdated integration of systems, lack of defined security practices, and billions of networked devices provide attack surfaces. Table 12.2 illustrates the cybersecurity risks of IIoT based on past studies.

Ultimately, IIoT cybersecurity issues include attacks, vulnerabilities, and threats. Advanced security, established processes, and proactive threat detection are needed to mitigate these threats. Industries can protect IIoT environments from growing cyberthreats and maintain operational continuity by employing strong countermeasures.

12.3.4 Integration of secure IIoT with AI

The combination of AI with the IIoT has the potential to greatly improve cybersecurity and operational efficiency. Serror et al. [36] show that integrating IIoT with AI enables real-time monitoring and decision-making, which improves system performance and resistance to cyberattacks. AI-powered solutions, especially machine learning (ML), analyze massive statistics created by IIoT devices to find abnormalities, improve operations, and forecast probable problems. Furthermore, incorporating blockchain technology into an AI-powered architecture creates a decentralized and immutable security layer, assuring the integrity and trustworthiness of IIoT networks. Collectively, AI and IIoT contribute to more secure, adaptable, and efficient smart manufacturing systems, allowing firms to reduce risks and improve their overall cybersecurity posture.

12.4 SECURE IIoT OPEN CHALLENGES AND POTENTIAL SOLUTIONS

Securing IIoT faces numerous has several obstacles. IIoT systems incorporate many heterogeneous devices with different security capabilities, making them complicated and huge. Ni and Li [37] highlight the difficulties of establishing consistent security across varied industrial contexts, especially with legacy technologies that were not built for current cybersecurity. Some IIoT devices' low processing capacity prevent them from using typical security procedures, rendering them susceptible to attacks. ML may improve IIoT security, although authors [38] note that training models require enormous datasets and AI systems can be attacked. Secure IIoT networks, which employ complex communication protocols, risk MITM attacks and data interception. IIoT devices may adapt to changing settings, needing flexible security solutions that can respond to new threats. IIoT's cloud and edge computing integration complicates managed and secured data storage and processing across platforms. IIoT ecosystems need machine intelligence, blockchain, and decentralized security.

Numerous IIoT security measures are needed to overcome several obstacles. Using lightweight encryption and authentication helps safeguard resource-constrained devices, while scalable security frameworks provide consistency across industrial contexts, including outdated systems [37]. ML may improve threat detection and response by evaluating real-time data for abnormalities, while federated learning reduces the requirement for centralized datasets and ML model training risks. Decentralized blockchain technology secures communication and data integrity, reducing MITM threats. Flexible, adaptive security measures should meet emerging threats and IIoT settings' dynamic nature [38]. Data stored and handled across platforms

Table 12.2 IIoT cybersecurity risks

Ref.	*Cybersecurity risks*		*Description*
[30]	*Attack*	*Botnet Attacks*	*Using compromised devices in a botnet to execute coordinated attacks.*
		Distributed Denial of Service (DDoS) Attacks	*Overloading IIoT networks to disrupt operations.*
[31]	*Vulnerability*	*Insecure Communication Protocols*	*Use of protocols without encryption or secure key exchange mechanisms.*
		Cloud Vulnerabilities	*Misconfigurations or weak access control in cloud services linked to IIoT.*
	Threats	*Lack of Standardization*	*Diverse device configurations and lack of unified security protocols increase vulnerabilities.*
[32]	*Threats*	*Insider Threats*	*Malicious actions or negligence by employees leading to security breaches.*
	Attack	*Ransomware Attack*	*Encrypting critical systems and demanding ransom to restore operations.*
		Command Injection	*Exploiting input vulnerabilities to execute unauthorized commands on devices.*
[33]	*Vulnerability*	*Insufficient Network Segmentation*	*Lack of proper isolation between IIoT devices, allowing attackers to move laterally.*
	Threats	*Data Breaches*	*Unauthorized access to sensitive industrial data, threatening confidentiality.*
		Espionage	*Stealing proprietary industrial data or intellectual property.*
	Attack	*Man-in-the-Middle (MITM) Attacks*	*Intercepting and manipulating communication between devices.*
[34]	*Vulnerability*	*Zero-Day Vulnerabilities*	*Exploits targeting previously unknown flaws in IIoT systems.*
	Threats	*Unauthorized Device Access*	*Exploiting weak authentication mechanisms to gain control over IIoT devices.*
	Attack	*Replay Attacks*	*Reusing legitimate network data to mimic authentic devices or users.*
		Privilege Escalation	*Gaining unauthorized access to critical system functionalities.*
[35]	*Vulnerability*	*Weak Authentication*	*Poor password policies or lack of biometric authentication.*
		Unpatched Firmware	*Outdated device firmware with exploitable vulnerabilities.*
		Limited Resource Devices	*Computationally weak devices incapable of running robust security protocols.*

may be secured by combining edge and cloud computing with strong security mechanisms. Machine intelligence, blockchain, and decentralized architectures may help enterprises build resilient IIoT ecosystems that adapt to new problems.

12.5 LEGAL REGULATORY GOVERNING SECURITY AND PRIVACY FOR IIoT

The IIoT raises security and privacy vulnerabilities that need substantial legal and regulatory frameworks to ensure safe and ethical adoption. In this study [39], the authors discovered that the heterogeneity and scale of IIoT networks made them susceptible to data breaches, unauthorized access, and system failures, necessitating special constraints. As stated by Gudlur et al. [40], AI-powered IoT devices provide extra privacy concerns about data collection, processing, and storage, necessitating GDPR and CCPA compliance. Bu et al. [41] emphasize the need of forensic preparation in IIoT situations, and advocate for traceable and auditable cyber incident investigation rules.

IIoT is safeguarded by global corporations and governments. The National Institute of Standards and Technology (NIST) Cybersecurity Framework focuses on cybersecurity risk, whereas ISO/IEC standards safeguard IIoT and IoT systems. These principles enhance data security, industrial safety, and breach accountability. Healthcare Health Insurance Portability and Accountability Act (HIPAA) and power North American Electric Reliability Corporation Critical Infrastructure Protection (NERC CIP) protect critical infrastructure. Secure IIoT ecosystem will be created by global standards and collaboration among governments, corporations, and technology providers.

12.6 CONCLUSIONS

In simple terms, considering their huge size, device heterogeneity, and essential nature, IIoT systems are both an urgent requirement and a challenging problem to secure. This chapter has shown how using AI-driven approaches like machine learning, blockchain, and edge computing may greatly improve IIoT security. Industries can guarantee operational resilience and safeguard sensitive data by addressing vulnerabilities and using adaptive security models. Moving ahead, sustained innovation, together with the adoption of uniform global standards, shall prove critical to mitigating increasing cybersecurity threats and fully realizing the promise for IIoT in Industry 4.0.

REFERENCES

1. Tyagi, A. K. (2024). Blockchain and Artificial Intelligence for Cyber Security in the Era of Internet of Things and Industrial Internet of Things Applications. In *AI and Blockchain Applications in Industrial Robotics* (pp. 171–199). IGI Global.
2. Hussain, Z., Akhunzada, A., Iqbal, J., Bibi, I., & Gani, A. (2021). Secure IIoT-enabled industry 4.0. *Sustainability*, 13(22), 12384.
3. Alotaibi, B. (2023). A survey on industrial Internet of Things security: Requirements, attacks, AI-based solutions, and edge computing opportunities. *Sensors*, 23(17), 7470.
4. Sasikumar, A., Ravi, L., Kotecha, K., Saini, J. R., Varadarajan, V., & Subramaniyaswamy, V. (2022). Sustainable smart industry: a secure and energy efficient consensus mechanism for artificial intelligence enabled industrial internet of things. *Computational intelligence and neuroscience*, 2022(1), 1419360.
5. Czeczot, G., Rojek, I., Mikołajewski, D., & Sangho, B. (2023). AI in IIoT management of cybersecurity for industry 4.0 and industry 5.0 purposes. *Electronics*, 12(18), 3800.
6. Jiang, J., Lin, C., Han, G., Abu-Mahfouz, A. M., Shah, S. B. H., & Martínez-García, M. (2022). How AI-enabled SDN technologies improve the security and functionality of industrial IoT network: Architectures, enabling technologies, and opportunities. *Digital Communications and Networks*, 9(6), 1351–1362.
7. Trakadas, P., Simoens, P., Gkonis, P., Sarakis, L., Angelopoulos, A., Ramallo-González, A. P., ... & Karkazis, P. (2020). An artificial intelligence-based collaboration approach in industrial iot manufacturing: Key concepts, architectural extensions and potential applications. *Sensors*, 20(19), 5480.
8. Lv, Z., Han, Y., Singh, A. K., Manogaran, G., & Lv, H. (2020). Trustworthiness in industrial IoT systems based on artificial intelligence. *IEEE Transactions on Industrial Informatics*, 17(2), 1496–1504.
9. Shahin, M., Maghanaki, M., Hosseinzadeh, A., & Chen, F. F. (2024). Advancing network security in industrial IoT: a deep dive into AI-enabled intrusion detection systems. *Advanced Engineering Informatics*, 62, 102685.
10. Yazdinejad, A., Kazemi, M., Parizi, R. M., Dehghantanha, A., & Karimipour, H. (2023). An ensemble deep learning model for cyber threat hunting in industrial internet of things. *Digital Communications and Networks*, 9(1), 101–110.
11. Latif, S., Driss, M., Boulila, W., Huma, Z. E., Jamal, S. S., Idrees, Z., & Ahmad, J. (2021). Deep learning for the industrial internet of things (iiot): A comprehensive survey of techniques, implementation frameworks, potential applications, and future directions. *Sensors*, 21(22), 7518.
12. Selvarajan, S., Srivastava, G., Khadidos, A. O., Khadidos, A. O., Baza, M., Alshehri, A., & Lin, J. C. W. (2023). An artificial intelligence lightweight blockchain security model for security and privacy in IIoT systems. *Journal of Cloud Computing*, 12(1), 38.
13. Demertzi, V., Demertzis, S., & Demertzis, K. (2023). An Overview of Privacy Dimensions on the Industrial Internet of Things (IIoT). *Algorithms*, 16(8), 378.
14. Rahman, A., Kundu, D., Debnath, T., Rahman, M., & Islam, M. J. (2024). Blockchain-based AI Methods for Managing Industrial IoT: Recent

Developments, *Integration Challenges and Opportunities. arXiv preprint arXiv:2405.12550.*
15. Zhang, F., Wang, H., Zhou, L., Xu, D., & Liu, L. (2023). A blockchain-based security and trust mechanism for AI-enabled IIoT systems. *Future Generation Computer Systems*, 146, 78–85.
16. Chen, J., Xue, J., Wang, Y., Huang, L., Baker, T., & Zhou, Z. (2023). Privacy-preserving and traceable federated learning for data sharing in industrial IoT applications. *Expert Systems with Applications*, 213, 119036.
17. Arachchige, P. C. M., Bertok, P., Khalil, I., Liu, D., Camtepe, S., & Atiquzzaman, M. (2020). A trustworthy privacy preserving framework for machine learning in industrial IoT systems. *IEEE Transactions on Industrial Informatics*, 16(9), 6092–6102.
18. Fu, A., Zhang, X., Xiong, N., Gao, Y., Wang, H., & Zhang, J. (2020). VFL: A verifiable federated learning with privacy-preserving for big data in industrial IoT. *IEEE Transactions on Industrial Informatics*, 18(5), 3316–3326.
19. Nguyen, D. C., Ding, M., Pathirana, P. N., Seneviratne, A., Li, J., Niyato, D., & Poor, H. V. (2021). Federated learning for industrial internet of things in future industries. *IEEE Wireless Communications*, 28(6), 192–199.
20. Ruzafa-Alcázar, P., Fernández-Saura, P., Mármol-Campos, E., González-Vidal, A., Hernández-Ramos, J. L., Bernal-Bernabe, J., & Skarmeta, A. F. (2021). Intrusion detection based on privacy-preserving federated learning for the industrial IoT. *IEEE Transactions on Industrial Informatics*, 19(2), 1145–1154.
21. Mukherjee, S., Gupta, S., Rawlley, O., & Jain, S. (2022). Leveraging big data analytics in 5G-enabled IoT and industrial IoT for the development of sustainable smart cities. *Transactions on Emerging Telecommunications Technologies*, 33(12), e4618.
22. Altamimi, S., Abu Al-Haija, Q. Maximizing intrusion detection efficiency for IoT networks using extreme learning machine. Discov Internet Things 4, 5 (2024). https://doi.org/10.1007/s43926-024-00060-x
23. Al-Haija, Q. A., Altamimi, S., AlWadi, M., Analysis of Extreme Learning Machines (ELMs) for intelligent intrusion detection systems: A survey, Expert Systems with Applications, Volume 253, 2024, 124317, ISSN 0957-4174, https://doi.org/10.1016/j.eswa.2024.124317
24. Malik, P. K., Sharma, R., Singh, R., Gehlot, A., Satapathy, S. C., Alnumay, W. S., ... & Nayak, J. (2021). Industrial Internet of Things and its applications in industry 4.0: State of the art. *Computer Communications*, 166, 125–139.
25. Alrawashdeh, K. & Al-Haija, Q. A. (2024). IIoT Security Using Industrial Identity Management System. In *NAECON 2024-IEEE National Aerospace and Electronics Conference*. 50–55. 10.1109/NAECON61878.2024.10670628
26. Pivoto, D. G., De Almeida, L. F., da Rosa Righi, R., Rodrigues, J. J., Lugli, A. B., & Alberti, A. M. (2021). Cyber-physical systems architectures for industrial internet of things applications in Industry 4.0: A literature review. *Journal of manufacturing systems*, 58, 176–192.
27. Khowaja, S. A., Dev, K., Qureshi, N. M. F., Khuwaja, P., & Foschini, L. (2022). Toward industrial private AI: A two-tier framework for data and model security. *IEEE Wireless Communications*, 29(2), 76–83.
28. Tan, S. F., & Samsudin, A. (2021). Recent technologies, security countermeasure and ongoing challenges of Industrial Internet of Things (IIoT): A survey. *Sensors*, 21(19), 6647.

29. Al-Haija, Q. A., & Al-Salameen, S. O. (2024). Biometric authentication system on mobile environment: A review. *Computer Systems Science & Engineering*, 48(4), 897–914.
30. Memos, V. A., Psannis, K. E., & Lv, Z. (2022). A secure network model against bot attacks in edge-enabled industrial Internet of Things. *IEEE Transactions on Industrial Informatics*, 18(11), 7998–8006.
31. Dhirani, L. L., Armstrong, E., & Newe, T. (2021). Industrial IoT, cyber threats, and standards landscape: Evaluation and roadmap. *Sensors*, 21(11), 3901.
32. Tsiknas, K., Taketzis, D., Demertzis, K., & Skianis, C. (2021). Cyber threats to industrial IoT: a survey on attacks and countermeasures. IoT, 2(1), 163–186.
33. Xenofontos, C., Zografopoulos, I., Konstantinou, C., Jolfaei, A., Khan, M. K., & Choo, K. K. R. (2021). Consumer, commercial, and industrial IoT (in) security: Attack taxonomy and case studies. *IEEE Internet of Things Journal*, 9(1), 199–221.
34. Jiang, X., Lora, M., & Chattopadhyay, S. (2020). An experimental analysis of security vulnerabilities in industrial IoT devices. *ACM Transactions on Internet Technology (TOIT)*, 20(2), 1–24.
35. Hajlaoui, R., Moulahi, T., Zidi, S., El Khediri, S., Alaya, B., & Zeadally, S. (2024). Towards smarter cyberthreats detection model for industrial Internet of Things (IIoT) 4.0. *Journal of Industrial Information Integration*, 39, 100595.
36. Serror, M., Hack, S., Henze, M., Schuba, M., & Wehrle, K. (2020). Challenges and opportunities in securing the industrial internet of things. *IEEE Transactions on Industrial Informatics*, 17(5), 2985–2996.
37. Ni, C., & Li, S. C. (2024). Machine learning enabled industrial iot security: Challenges, trends and solutions. *Journal of Industrial Information Integration*, 38, 100549.
38. Ashok, K., Boddu, R., Ali Syed, S., Sonawane, V. R., Dabhade, R. G., & Shaker Reddy, P. C. (2023). GAN Base feedback analysis system for industrial IOT networks. *Automatika: časopis za automatiku, mjerenje, elektroniku, računarstvo i komunikacije*, 64(2), 259–267.
39. Saif, A. & Al-Haija, Q. A. (2024). *Artificial Intelligence (AI)-Powered Internet of Things (IoT)*. 10.1201/9781032648309-3
40. Gudlur, V. V. R., Shanmugan, V. A., Perumal, S., & Mohammed, R. M. S. R. (2020). Industrial internet of things (IIoT) of forensic and vulnerabilities. *International Journal of Recent Technology and Engineering*, 8(5), 257–260.
41. Bu, L., Zhang, Y., Liu, H., Yuan, X., Guo, J., & Han, S. (2021). An IIoT-driven and AI-enabled framework for smart manufacturing system based on three-terminal collaborative platform. *Advanced Engineering Informatics*, 50, 101370.

Chapter 13

The AI shield

Enhancing the security in industrial IoT

Vasavi Sravanthi Balusa, Kishor Kumar Reddy C, Harika Koormala, Ch Rajyalakshmi, and Srinath Doss

13.1 INTRODUCTION

The Industrial Internet of Things (IIoT) represents a specialized subset of the Internet of Things (IoT) that enhances industrial sectors such as manufacturing, energy, healthcare, transportation, and agriculture. It achieves this by interconnecting devices and systems, facilitating advanced automation, real-time data analysis, and improved operational efficiency. While IIoT enhances productivity and safety, its expansive networks and reliance on interconnected devices introduce unique security challenges. Industrial Internet of Things (IIoT) systems, frequently situated in remote or susceptible locations, are increasingly susceptible to cyber threats owing to the essential nature of industrial data. Securing these systems presents greater complexity than traditional IT environments due to the integration of various devices, legacy systems, and physical assets. This complexity necessitates advanced solutions capable of adapting to the evolving cyber threat landscape. One essential tool for tackling IIoT security issues is artificial intelligence (AI). By leveraging AI's capabilities in predictive analytics, real-time threat detection, and anomaly identification, industries can bolster their defenses against both digital and physical vulnerabilities. Techniques like edge computing further enhance data privacy and real-time decision-making, while federated learning addresses privacy concerns by enabling decentralized data analysis. But there are drawbacks to integrating AI with IIoT as well, such as resource limitations on edge devices, concerns about data privacy, and the moral ramifications of automated decision-making. By enabling stronger cryptography and quicker threat detection, emerging technologies such as quantum computing hold the potential to dramatically transform IIoT security and open the door to a more robust industrial environment.

An obscure subset of the Internet of Things focused on industrial applications, the Industrial Internet of Things (IIoT) uses networked devices and systems to improve manufacturing, energy, healthcare, transportation, and

 DOI: 10.1201/9781003606307-13

agricultural operations. IIoT increases efficiency, productivity, safety, and flexibility by gathering and analyzing real-time data from machines and sensors. It involves integrating sensors, actuators, and controllers with industrial equipment to enable data collection, optimized control, and automation. Key components include smart sensors, edge devices, gateways, cloud platforms, analytics software, and security frameworks, which together ensure efficient data handling. In manufacturing, IIoT powers smart factories, providing insights for predictive maintenance, quality monitoring, and production automation. The healthcare mechanism is changing: an entire country of tradition being challenged by ethical and legal concerns regarding healthcare quality, excessive work restrictions, surgical procedure costs, and their repercussions [1]. In energy, IIoT enhances efficiency and environmental impact by monitoring grids, wind turbines, and pipelines. In healthcare, IIoT aids patient monitoring and asset tracking through medical devices and wearables. In transportation, IIoT supports fleet management and route optimization, improving logistics and reducing costs. In agriculture, it monitors soil, livestock, and automates processes, optimizing yields and sustainability. IIoT enhances productivity by automating tasks, supports predictive maintenance to reduce downtime, improves safety by monitoring hazards, and enables innovation through data insights. However, challenges include interoperability among varied devices, massive data management needs, security vulnerabilities, and scalability issues. Complex network architecture and a lack of network segmentation increase risks, as does insufficient monitoring. Data privacy is a concern due to the large amounts of sensitive information generated and transmitted. Physical security is limited for devices in remote or exposed locations. It enables real-time threat detection and response, recognizing network anomalies and suspicious behaviors by using encryption and security protocols [2]. AI solutions efficiently monitor large-scale IIoT networks and provide automated threat intelligence. By integrating edge computing, AI can detect threats locally, reducing latency. Predictive analytics help anticipate and prevent issues, with AI-driven risk assessments that identify and prioritize vulnerabilities. AI improves endpoint security through device authentication [3], intrusion detection, and zero-trust models. These models are illustrated in Figure 13.1.

It enhances data protection with automated encryption and data integrity checks, ensuring privacy compliance. AI also automates security operations, reducing human error, and supports rapid incident response. AI addresses evolving threats with adaptable learning, helping detect previously unseen attack patterns. It assists in threat hunting and post-incident forensics, identifying vulnerabilities and strengthening defenses. By integrating AI, IIoT systems gain scalable, real-time protection, predictive security, and robust data privacy, making industrial ecosystems more resilient against security challenges as IIoT adoption expands.

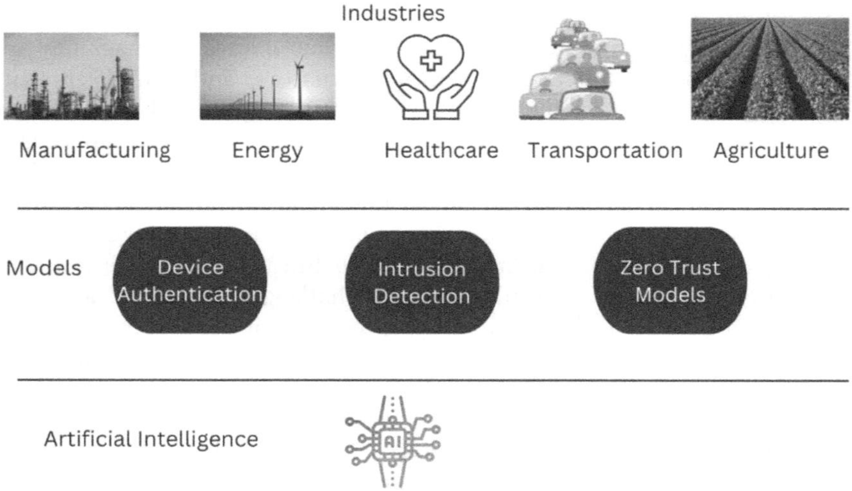

Figure 13.1 AI models for securing IIoT.

13.2 HAZARD ENVIRONMENT IN THE INDUSTRIAL IoT

Industrial IoT (IIoT) environments face various security threats targeting both digital and physical systems, posing significant risks to industrial operations. Key threats include cyber-physical attacks, network breaches, and vulnerabilities in devices, communications, data privacy, and architectural design. Cyber-physical attacks exploit both digital and physical elements of IIoT, often leading to real-world damage. For instance, the Stuxnet worm manipulated nuclear centrifuges [4, 5], and the Triton attack disabled safety systems, jeopardizing human safety. Such attacks can damage equipment, disrupt operations, and compromise safety. Network breaches in IIoT often involve data stealing, ransomware, and also DDoS attacks. Device-level attacks exploit firmware vulnerabilities, especially when IIoT devices run outdated software, enabling attackers to access networks or spread malware. Physical tampering is a risk for remote or unsupervised IIoT devices, where attackers can spoof identities or install rogue hardware. Compromised devices can become part of botnets, which attackers control to launch DDoS or malware distribution.

Man-in-the-middle (MITM) attacks, which compromise data and control, enable attackers to halt and alter data using IIoT transmission protocols like MQTT and CoAP, which are frequently unsecure. Data privacy and integrity threats arise from unencrypted data transmission, where attackers can intercept sensitive information or alter data to create operational hazards. Denial of Service (DoS) attacks disrupts critical systems, either through broad outages or selective targeting of essential components, leading to significant operational challenges. The IIoT architecture introduces unique vulnerabilities. Device-level weaknesses [6] stem from limited processing power, weak authentication, outdated firmware, and inadequate security features. Network vulnerabilities include poor segmentation, insecure protocols, and lack of real-time monitoring, allowing lateral movement by attackers within networks. Data security risks involve insecure data storage, insufficient encryption, and inconsistent access controls. Cloud and edge computing in IIoT are prone to misconfiguration, insufficient edge security, and risks across hybrid architectures, where each layer's interactions create security gaps. Integrating edge computing, cloud, and IoT reduce the quantity of cloud-IoT application failures [7].

Integrating legacy systems with IIoT introduces compatibility and update challenges, while third-party components bring supply chain vulnerabilities that can compromise security before devices are deployed. Physical security remains an issue for IIoT devices in remote locations, where tampering risks are high, and harsh environments may affect device functionality. Human factors play a role, as lack of security training leads to risky behaviors, and configuration errors create exploitable weaknesses. Misuse of privileged access can result in data breaches or sabotage. Ensuring security across all levels of IIoT architecture is essential to protect against these complex threats, secure operations, and safeguard sensitive data in connected industrial environments.

13.3 ARTIFICIAL INTELLIGENCE APPROACHES FOR IIoT SECURITY

Manufacturing is being transformed by the Industrial Internet of Things (IIoT), which combines sensors, devices, machines, and systems to collect and analyze data. This shift offers enhanced efficiency, predictive maintenance, and smarter decision-making. By identifying threats, identifying abnormalities, and authorizing adaptive defense systems, machine learning (ML) significantly contributes to IIoT security. ML analyzes vast amounts of data in real time to locate patterns that are difficult to find manually. This recommends both proactive and reactive security for IIoT systems. IIoT systems are required in productions like manufacturing, energy, healthcare, and transportation, where security violations or downtime can have severe outcomes. Key security challenges include a vast attack surface, real-time operations, and legacy devices lacking security, and data integrity and privacy. ML systems can adapt to evolving threats in real-time, making them effective against dynamic attack patterns [8]. Anomaly detection, intrusion detection and prevention, predictive maintenance, threat intelligence, automated incident response, access control, and data protection are just a few of the ways machine learning (ML) improves IIoT security. These are illustrated in Figure 13.2.

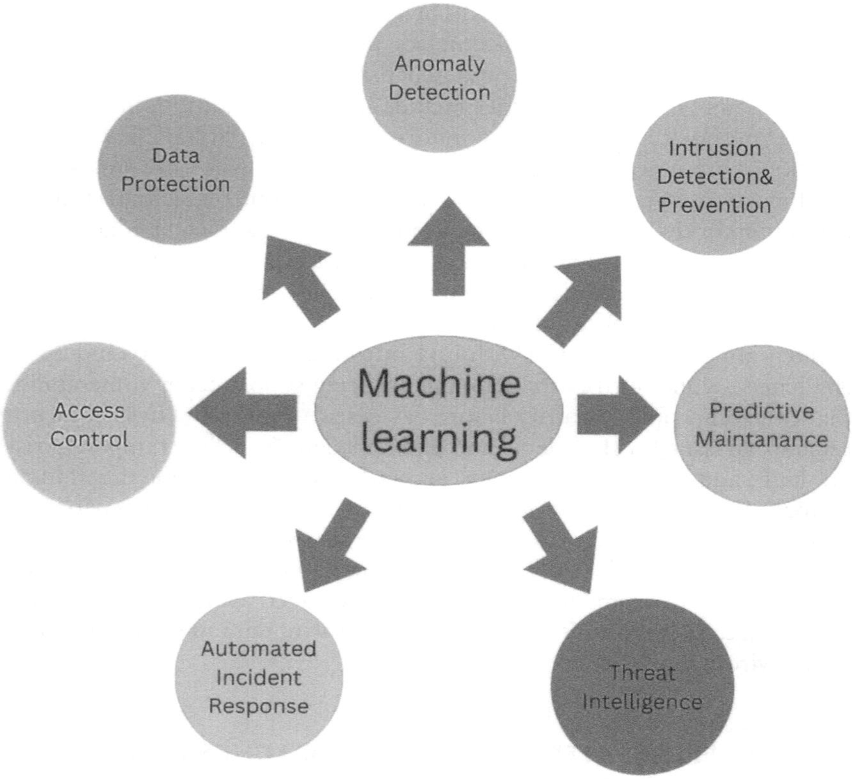

Figure 13.2 ML methods to enhance IIoT security.

ML detects unusual device behavior, analyzes network traffic for cyberattacks, and monitors environmental changes to signal hardware issues. Intrusion detection and prevention systems (IDPS) in IIoT benefit from ML by automatically identifying threats based on traffic patterns and applying corrective actions. Machine learning (ML) methods provide advantages in finding and stopping a wide range of intrusions, including known and unknown threats [9].

Predictive maintenance uses ML to predict equipment malfunctions, minimizing the risk of attacks by reducing equipment failure. ML also supports threat intelligence, providing insights into evolving attack patterns, classifying threats, and mapping potential vulnerabilities. Automated incident response enables faster threat mitigation, while adaptive defenses ensure systems can respond to evolving attack strategies without extensive human intervention. Access control is enhanced by behavioral biometrics and risk-based authentication, while ML optimizes data encryption and privacy, securing sensitive data in transmission. Implementing ML in IIoT security

Table 13.1 AI's role in securing IIoT

AI application	*Description*	*Impact*
Predictive Security	AI uses machine learning models to predict threats before they happen by analyzing historical data.	Shifts IIoT security from reactive to proactive, reducing the likelihood of breaches and system downtime.
Autonomous Cybersecurity Systems	AI-powered systems recognize, evaluate, and react to threats on their own in real time.	reduces the need for manual security management operations by improving responsiveness and agility.
Federated Learning for Data Privacy	AI allows decentralized learning across devices without sharing sensitive data, ensuring privacy.	Protects sensitive data while still enabling collective threat intelligence across IIoT networks.
Quantum AI for Enhanced Security	Quantum computing powers AI to process vast datasets quickly and create quantum-resistant encryption models.	Provides advanced threat detection and encryption models to protect against quantum-based attacks.
AI-Driven Risk Management	AI evaluates and prioritizes security risks based on real-time data, optimizing resource allocation.	Helps organizations focus resources on the highest priority threats and improves compliance with regulations.
Collaboration and Threat Intelligence	AI facilitates shared threat intelligence across IIoT networks, enabling coordinated defense efforts.	Strengthens global IIoT security by sharing threat data and responses to neutralize global threats faster.

comes with challenges, such as data quality, scalability, managing false positives and negatives, model explainability, and keeping up with evolving threats [10].

Table 13.1 outlines the AI applications and its description with an impact in providing security on IIoT.

Deep learning technologies have huge potential for improving intrusion detection in IoT environments [11]. Chain optimization, energy management, automation, environmental monitoring, real-time decision-making, and supply chain security are critical components of a smart, efficient, and sustainable industrial ecosystem. DL enables predictive maintenance by analyzing sensor data for potential failures, reducing downtime and extending equipment life. In quality control, by automating defect detection, inspecting products with high accuracy, and enabling autonomous inspection, it optimizes supply chains by forecasting demand, optimizing routes, and managing inventory. Automation and robotics powered by DL improve IIoT innovation. Robotics plays an important role in this transformation by giving substantial capabilities that drive automation, efficiency, and flexibility in production processes [12]. Robots' strengths in automation, accuracy, teamwork, and adaptability greatly increase operational effectiveness, save

Table 13.2 AI techniques for enhancing IIoT security

Technique	*Application in IIoT security*	*Key benefit*
Machine Learning (ML)	Examines enormous datasets to find irregularities and forecast any security threats.	Provides predictive security that can foresee threats before they manifest.
Deep Learning (DL)	Identifies complex patterns and behaviors across IIoT systems.	Increases threat detection accuracy by understanding complex data relationships.
Reinforcement Learning (RL)	Enables systems to learn optimal security strategies through trial and error.	Enhances autonomous decision-making and real-time security responses.
Federated Learning	Decentralized learning without data sharing among IIoT devices.	Protects privacy and complies with data security regulations.
Quantum Computing	Enhances AI's processing power to solve complex problems and improve security.	Enables faster detection and creation of quantum-resistant encryption.

costs, increase safety, and help companies meet the ever-evolving demands of the market [5]. Robots with DL-powered vision systems perform complex tasks on assembly lines, while autonomous mobile robots (AMRs) use DL to navigate efficiently.

Table 13.2 highlights AI techniques and key benefits for enhancing IIOT security

DL also enhances safety in IIoT by detecting gas leaks, monitoring hazardous materials, and ensuring worker safety.

13.4 DETECTION OF ANOMALIES AND INTRUSIONS IN INDUSTRIAL IoT

Intrusion Detection Systems (IDS) are crucial for network security since they were designed to keep an eye on system activity and network traffic for signs of malicious activity, illegal access, or possible assaults. Hybrid approach, which combines deep learning models with rule-based features selection techniques, increases detection accuracy, decreases computational complexity, and ensures efficient anomaly detection tailored for resource-constrained IIoT environments [13]. There are various types of IDS based on detection methods, deployment architectures, and the level of data monitoring.

Network traffic is examined by a Network-Based Intrusion Detection System (NIDS) to look for unknown behavior or known attack signatures. Installed on different devices, such as workstations or servers, Host-Based Intrusion Detection Systems (HIDS) [14] identify host-level activity. Hybrid IDS unite the advantages of both NIDS and HIDS for a wider approach to intrusion awareness. By comparing incoming data to a database of attack signatures, Signature-Based Intrusion Detection Systems (SIDS) identify known threats. An assault may be identified using anomaly-based intrusion detection systems (IDS), which are essential for protecting Industrial IoT (IIoT) environments against complex and dynamic cyberthreats [15]. These systems look for departures from a baseline of typical behavior. Behavior-Based Intrusion Detection Systems (BIDS) concern on the behavioral patterns of users, processes, and devices.

Anomaly awareness is the process of identifying methods in data that do not accept to expected behavior. It is essential in industries such as cybersecurity, finance, healthcare, and industrial systems to detect hazards, fraud, system failures, or unusual behaviors. Conventional anomaly monitoring techniques, including statistical or rule-based systems, frequently face competition from vast amounts of data, intricate patterns, and dynamic threats. Anomaly detection is improved by artificial intelligence (AI), especially machine learning (ML) and deep learning (DL), which enable computers to learn from data without being explicitly taught to identify specific anomalies. AI-based methods are more flexible and efficient, capable of observing complicated patterns in large datasets. Supervised anomaly observation involves training a model on labelled data, where normal and anomalous ideals are provided. Support Vector Machines (SVM), Logistic Regression, and Decision Trees are examples of classification models that are frequently employed. By learning from network traffic data and differentiating between benign and malevolent behavior, SVM has demonstrated significant potential in enhancing the accuracy and effectiveness of IDS [16]. Unsupervised anomaly observation methods, in contrast, do not require labelled data. Auto encoders, which employ neural networks to rebuild data and detect anomalies based on reconstruction errors, and clustering-based approaches, such as K-means and DBSCAN, are examples of common techniques. Anomaly detection in high-dimensional, complicated datasets is accomplished using Deep Learning techniques such as Deep Auto-encoders, Convolutional Neural Networks (CNNs), and Recurrent Neural Networks (RNNs) [17]. Reinforcement Learning (RL) offers an innovative reach to anomaly detection, permitting systems to dynamically adjust detection parameters on the basis of feedback from their environment. AI-based solutions are essential for overcoming obstacles such the large amount of data, legacy systems, and changing cyberthreats in real-time threat detection for Industrial Internet of Things (IIoT) environments.

13.5 LEVERAGING AI IN INDUSTRIAL IoT TO IDENTIFY HAZARDS

Predictive logical is a strong tool that uses analytical algorithms, machine learning, and data mining methods to estimate future outcomes by analyzing and historical and real-time data. In cybersecurity, its primary target

Table 13.3 Benefits of AI in IIoT security

Benefit	*Description*	*Impact on IIoT security*
Scalability	AI systems scale effortlessly with growing IIoT networks.	Ensures security measures and remain effective as IIoT ecosystems expand.
Real-Time Threat Response	AI analyzes and responds to threats instantaneously.	Improves the ability to contain and mitigate threats in real time.
Continuous Improvement	AI models evolve and get smarter over time with exposure to new data.	Enables IIoT security to stay ahead of emerging threats.
Automated Incident Response	AI automates response to security incidents, reducing manual intervention.	Reduces downtime and minimizes potential damage during attacks.
Proactive Threat Detection	AI can detect potential threats before they fully develop.	Shifts focus to preventing security breaches rather than recovering from them.

is to forecast potential security hazards and vulnerabilities before they exist, allowing organizations to take cautious measures to relieve risks. Organizations can improve their security posture and lower the risk of cyberattacks, breaches, or system failures by using predictive analytics, which analyzes data from various sources such as network traffic, system logs, user behavior, and external threat intelligence.

Table 13.3 provides an overview of various benefits of AI in IIoT.

Predictive maintenance manufacturers, years of manufacturing, and machine kinds are among the specific data that IoT gathers. The system then uses these statistics to predict maintenance [18]. A procedure of predictive analytics in security involves collecting data from various sources, pre-processing it to check accuracy, and identifying essential features that influence security threats. Subsequently, machine learning algorithms are trained using historical data. The trained model can forecast future security threats, such as prospective cyberattacks or system vulnerabilities. Common methods used in predictive security systems contain anomaly detection [19], predictive modelling, behavioral analytics, and threat intelligence integration. In real-world implementations, predictive analytics is used over industries like finance, healthcare, and IIoT. In the financial sector, it helps detect and intercept fraud, while in healthcare, it discovers medical device behavior to detect possibilities cyberattacks. In IIoT, predictive analytics enhances the protection of industrial control systems (ICS) and critical infrastructure, as well as assists in predictive conservation. In cybersecurity operations centers (SOC), prophetic analytics aids in threat hunting and incident answer by providing insights into appearing threats and prioritizing incidents based on risk.

AI models can process huge amounts of data, identify patterns, and make predictions, helping organizations predict security threats, assess risk levels, and take cautious countermeasures. These artificial intelligence models improve cybersecurity defenses and the capacity to keep ahead of changing threats. Because AI-powered systems can predict the likelihood of security events and identify new threats before they materialize, they are essential to risk assessment and threat forecasting. Unlike traditional methods that rely on past incidents, several AI models are developed in these processes, each contributing unique strengths. Supervised learning models are commonly used for risk assessment and threat forecasting. Unsupervised learning models, which do not prefer labelled data, are useful for observing unknown hazards and vulnerabilities by estimating hidden patterns or anomalies in huge datasets. Reinforcement learning (RL) is important for dynamic environments like cybersecurity, where models modify and develop based on feedback from the environment. In risk assessment, RL models continuously adjust security strategies based on the occurrence of cyberattacks. In threat forecasting, RL models optimize responses to security threats, learning the best course of action to mitigate risks.

Deep researching models, which use multi-layered neural networks to learn complex patterns, are specifically constructive for analyzing high-dimensional or unstructured data. Convolutional neural networks (CNNs) can detect attacks by inspecting raw packet data or system logs. Time-series data from security logs and other sequential data can be analyzed by recurrent neural networks (RNNs), especially Long Short-Term Memory (LSTM) networks, which are effective tools for predicting future assaults based on historical patterns. AI models have diverse applications across industries. In the financial sector, AI models are used for cheating detection, credit risk assessment, and market estimating. These models inspect transaction data, financial behavior, and market trends to forecast and mitigate financial threats. AI models are used in healthcare to preserve patient privacy and safety by keeping an eye out for cybersecurity threats on medical devices and forecasting risks associated with data breaches. By calculating risks in industrial control systems (ICS) and assisting with predictive maintenance, artificial intelligence (AI) models in the Industrial Internet of Things (IIoT) help avoid equipment failures that could result in security flaws. These AI models are illustrated in Figure 13.3.

In cybersecurity operations, AI models integrate into Security Operations Centers (SOC) to improve threat detection and response. Threat intelligence tools driven by AI examine global threat data and offer practical insights into new threats. Overall, AI models significantly enhance the ability to assess risks, forecast threats, and implement proactive security measures across various industries, making them indispensable tools in modern cybersecurity.

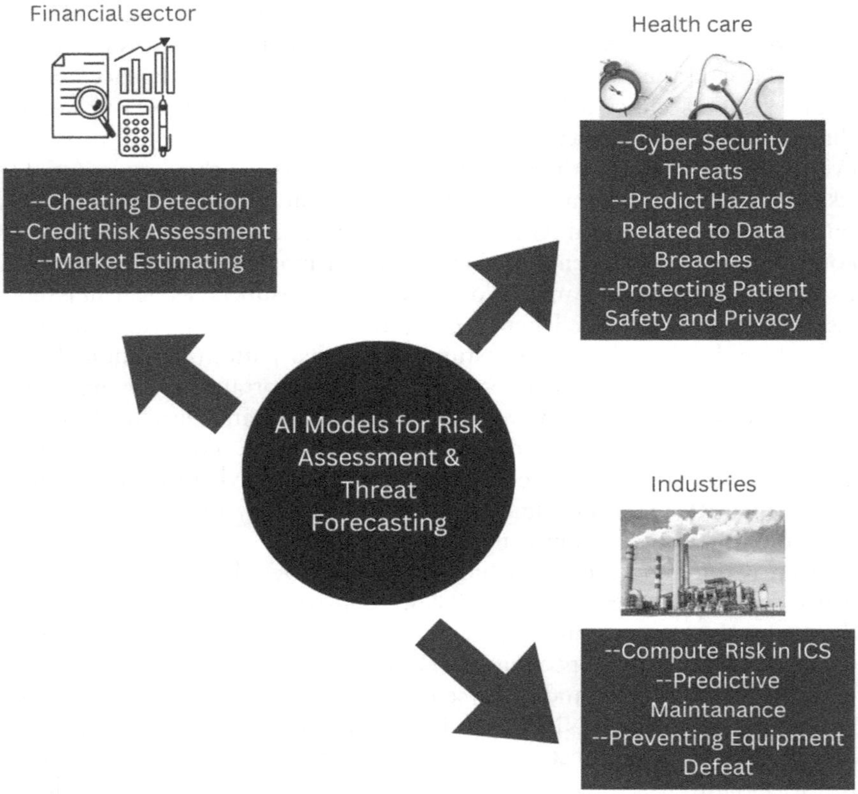

Figure 13.3 AI models for risk assessment and threat forecasting.

13.6 AI-POWERED NETWORK AND ENDPOINT SECURITY FOR INDUSTRIAL IoT

Because sectors such as manufacturing, shipping, energy, and healthcare are still embracing the Industrial Internet of Things (IIoT), it is becoming increasingly vital to secure IIoT devices. Although there are many obstacles to overcome, protecting IIoT devices is essential to avoiding problems, safety risks, and large financial losses. Artificial Intelligence (AI) contributes significantly to cybersecurity calculations by offering quick, flexible, and intelligent defenses to protect these systems from changing threats. Artificial Intelligence (AI) contributes significantly to cybersecurity calculations by offering quick, flexible, and intelligent defenses to protect these systems from changing threats.

Table 13.4 shows the role of AI functionality in IIOT security management.

AI-driven solutions are key to overcoming these challenges. Machine learning algorithms can create behavioral baselines for devices, detecting

Table 13.4 AI-Driven security management in IIoT

AI functionality	*Role in IIoT security management*	*Key outcome*
Dynamic Risk Assessment	AI evaluates risk in real-time based on current data to adjust security protocols.	Continuously adjusts security measures based on risk levels.
Prioritization of Resources	AI allocates resources to the most pressing threats, optimizing security efforts.	Ensures efficient resource use, focusing on critical security needs.
Automated Compliance Monitoring	AI ensures that IIoT systems are adhering to relevant industry regulations.	Minimizes compliance risks and prevents costly fines.

anomalous activities in real time. AI is already being applied across various industries to secure IIoT devices. In manufacturing, AI detects equipment anomalies that could indicate an early-stage attack or system failure. In the energy sector, AI monitors grid stability and detects tampering to prevent operational disruptions. In healthcare, AI secures medical IoT devices by preventing unauthorized access and ensuring patient data safety. Looking ahead, AI's role in IIoT security will continue to evolve. Adaptive security frameworks will learn from each incident, enhancing defenses. Collaboration with edge computing will enable faster, localized threat detection and response while reducing bandwidth and latency. Industry-wide collaboration to establish AI-based security standards will further strengthen the security of IIoT systems across sectors. AI-driven solutions are critical for addressing the unique security challenges of IIoT, enabling intelligent detection, rapid response, and scalable defense mechanisms. Network security is crucial to safeguarding sensitive information, confirming business continuity, and maintaining trust in an interconnected world. Intrusion Prevention Systems (IPS) and Threat Intelligence (TI) are important instruments in obtaining networks by detecting and neutralizing cyber risks. An IPS monitors network activity and takes action to inspect intrusions. It can be network-based (NIPS), which observes traffic across the network, or host-based (HIPS), which is deployed on individual devices to protect endpoints. Core functions of IPS include threat detection, prevention, policy enforcement, and logging for forensic analysis. Threat intelligence is an important security technique to understand cyber-risks [20]. It includes strategic, operational, tactical, and technical data to provide insights into attack vectors, tactics, and vulnerabilities. By leveraging the latest threat data, organizations can detect both known and unknown threats faster and automate the response to mitigate emerging risks. AI-driven behavioral analysis and anomaly detection further optimize these systems by identifying new and advanced threats.

Combining IPS with threat intelligence provides several benefits. It permits faster incident reaction by automating risk detection and mitigation, decreases false positives by improving alert accuracy, and ensures real-time, adaptive defenses. This integration creates a comprehensive risk management approach by continuously identifying vulnerabilities and proactively addressing threats. In finance, these tools help prevent fraud and ransomware by monitoring malicious IPs and domains. In healthcare, they secure patient data and medical devices from violations. Governments use these systems to monitor nation-state actors and APTs, while e-commerce businesses defend against attacks targeting customer data. Together, IPS and threat intelligence provide a powerful, adaptive defense against evolving cyber threats. By proactively identifying and mitigating risks, these tools are essential to safeguarding networks in an increasingly digital world.

13.7 DATA SECURITY AND PRIVACY IN INDUSTRIAL IoT USING AI

Securing Industrial Internet of Things (IIoT) systems needs dependable data relaying and encrypted data. Encryption ensures the confidentiality, integrity, and safety of data against cyber threats when sensitive data is transferred between devices, networks, and cloud services [21].

Table 13.5 highlights AI-powered IIoT security technologies with their role and benefits.

Best practices for securing IIoT data include implementing strong encryption standards, ensuring end-to-end encryption, and regularly updating firmware. AI can also enhance privacy through advanced techniques such as federated learning, which keeps data localized on devices, and differential privacy, which introduces noise to data to preserve individual privacy. Homomorphic encryption allows AI algorithms to analyze encrypted data without exposing it, providing secure processing capabilities. By engaging a merging of encryption, secure transmission protocols, and AI-driven

Table 13.5 AI-Powered IIoT security technologies

Technology	*Role in IIoT security*	*Benefit*
Machinel Learning	Analyzes network behavior to detect anomalies and predict attacks.	Proactively addresses threats before they escalate.
Federated Learning	Allows decentralized learning while maintaining data privacy.	Enhances privacy and security without sharing sensitive data.
Quantum AI	Uses quantum computing to enhance AI's processing power and security.	Offers better encryption and threat detection capabilities.
Block chain	Ensures data integrity and transparency in IIoT transactions.	Enhances the trustworthiness of IIoT data and processes.

techniques, IIoT systems can preserve data integrity, protect privacy, and protect against evolving cybersecurity hazards, structured updates, compliance with privacy regulations, and continuous monitoring is essential for preserving robust data defense in IIoT resources. Edge computing is a transformative technology that enables data processing, analysis, and decision-making near the source of data generation, typically at the "edge" of the network. Two state-of-the-art CNN architectures, ConvNeXt and ResNet152V2, enhance intrusion detection in edge computing environments [22]. By enabling local data processing within IIoT devices, gateways, and nearby nodes, edge computing reduces dependency on centralized data centers and cloud platforms in the context of the Industrial Internet of Things (IIoT). This reach offers numerous benefits, involving reduced latency, enhanced data privacy, improved reliability, and efficient resource usage. It empowers local data processing in IIoT, driving industrial operations forward. Reduced latency, improved data privacy and security, optimum bandwidth use, increased reliability, and scalability are some of the main advantages of edge computing for IIoT. These benefits are shown in Figure 13.4.

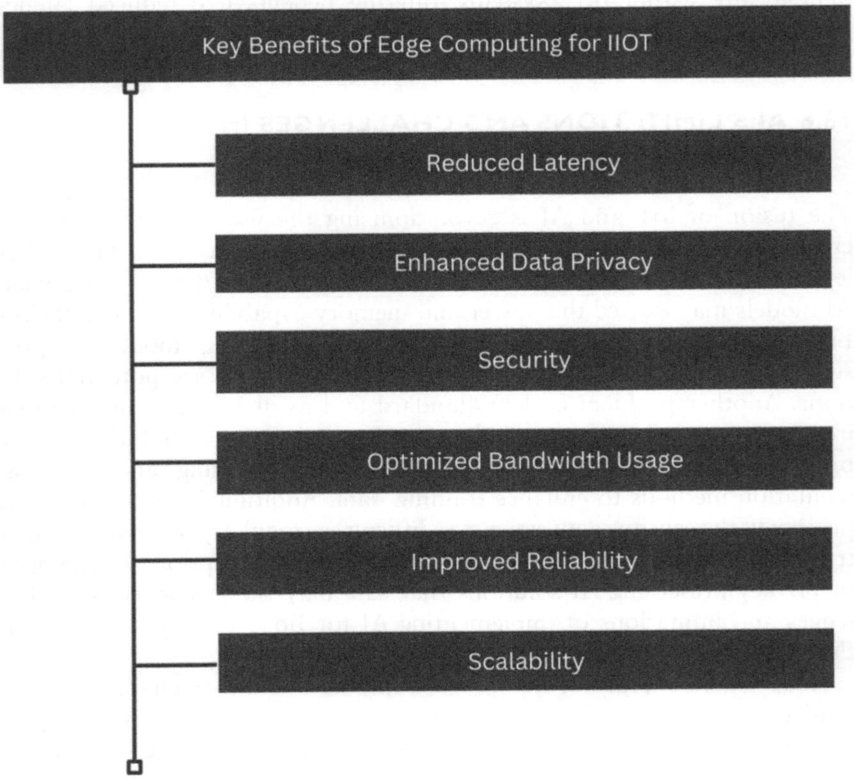

Figure 13.4 Key benefits of edge computing for IIoT.

Edge computing increases data security by reducing vulnerability to hackers by processing critical data locally. By minimizing bandwidth and sending only pertinent data to the cloud, it reduces network congestion. Its design consists of cloud layers for sophisticated storage and analysis, device layers for data gathering, edge gateways for processing, and optional fog layers for near-edge capabilities. In IIoT environments, edge computing makes real-time applications possible, such as automated decision-making, quality control, and predictive maintenance. It guarantees product quality by identifying defects, minimizes downtime by identifying equipment problems early, and facilitates autonomous control for vital sectors including the automobile and energy industries. This improves safety and operating efficiency. Edge AI increases edge computing by uniting artificial intelligence with local data processing. To successfully implement edge computing in IIoT, it is important to select appropriate edge hardware, use AI frameworks like TensorFlow Lite and OpenVINO, implement robust security protocols, and optimize data processing by filtering and compressing data locally. Regularly updating edge AI models ensures accuracy and resilience against evolving challenges. In conclusion, edge computing is a powerful approach for processing data locally within IIoT systems, offering benefits like reduced latency, enhanced privacy, and improved reliability.

13.8 AI'S LIMITATIONS AND CHALLENGES IN PROTECTING IIoT

The fusion of IoT and AI is revolutionizing the way industries operate, enhancing productivity, operational efficiency, and safety [23]. These challenges include limited computational resources on edge devices [24], where AI models may exceed the power and memory capabilities of IIoT devices. Lightweight AI models optimized for edge computing, model compression techniques, and specialized hardware accelerators are potential solutions. Another problem is data standard and availability, as inconsistent, incomplete, or noisy data can influence AI model accuracy. Solutions to this problem include data pre-processing, using clear learning, and data augmentation methods to enhance training data. Additionally, balancing data privacy with security requirements is difficult, particularly in sensitive industries. Federated learning, differential privacy, and encryption are examples of privacy-preserving AI solutions that can allay these worries. The challenges and limitations of implementing AI for IIoT security are shown in Figure 13.5.

Adapting to rapidly evolving cyber threats is another challenge, as AI models trained on historical data may struggle with novel attacks. Continuous learning models, hybrid models that combine rule-based detection with AI, and regular model updates can help mitigate this. Interoperability and standardization obstacles arise due to various device protocols and

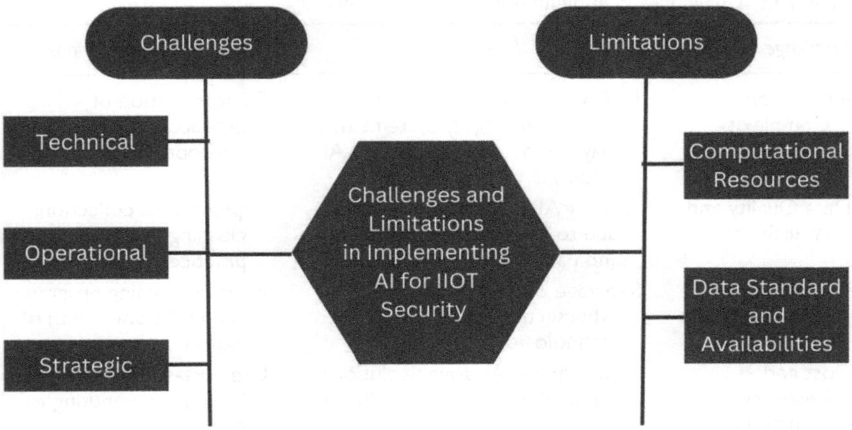

Figure 13.5 Challenges and limitations of implementing AI for IIoT security.

hardware platforms. Modular AI solutions, middleware, and industry standards can facilitate compatibility and deployment. Additionally, the restricted availability of skilled personnel in AI and cybersecurity, together with the high costs of AI implementation, can obstruct adoption. Solutions involve upskilling staff, using automated machine learning tools, and adopting phased AI implementations to decrease costs. Data science and AI development highlight the significance of ethical issues in modern technologies [25].

Ethical thoughts also appear regarding data privacy, algorithmic bias, fairness, and transparency [26]. Addressing these treats requires transparency in data gathering, informed consent, and the use of typical datasets to neglect biased AI models. By addressing these challenges and ethical concerns, organizations can effectively leverage AI for IIoT security, improving real-time threat detection, operational efficiency, and data protection while ensuring that AI systems remain ethical, transparent, and trustworthy. Although artificial intelligence (AI) offers powerful tools to improve the security, effectiveness, and operational insights of Industrial Internet of Things (IIoT) systems, it has been established that AI models have significant limitations when used in these complex situations. These restrictions arise from the constraints of IIoT architectures, the challenging requirements of industrial applications, and the evolving cyber hazards targeting IIoT systems. To improve AI's resilience, scalability, and applicability in IIoT, these issues must be resolved.

Table 13.6 Summarizes the challenges and potential solutions in implementing AI for IIOT security.

Transitioning to Industry [27] is a journey that requires an in-depth rethinking of existing processes, strategies, and organizational culture. A large amount of processing power and memory are needed for many AI models, especially deep learning models, which are computationally

Table 13.6 Challenges in implementing AI for IIoT security

Challenge	*Description*	*Potential solutions*
Integration Complexity	IIoT networks include diverse devices and legacy systems that may not be compatible with AI solutions.	Standardization of protocols and improved interoperability.
Data Quality and Availability	Training AI models is challenging due to the unstructured, loud, and partial nature of IIoT data.	Improve data collection, cleaning, and management practices.
Cybersecurity Talent Shortage	Shortage of experts in both cybersecurity and AI technologies.	Invest in training programs and collaborate with AI experts.
Cost and Resource Constraints	Implementing AI-driven solutions can be expensive, especially for SMEs.	Use cost-effective solutions like edge computing to minimize costs.

demanding. However, IIoT edge devices often have limited hardware capabilities, operating with minimal storage, processing power, and energy resources. Developing lightweight models for edge computing, employing privacy-preserving techniques, improving data pre-processing methods, and adopting explainable AI frameworks [28]. Data imbalance is a common challenge, as some IIoT systems generate limited or biased data, making training models difficult. Sensor crashes, communication issues, and environmental interference can occur in missing, corrupted, or noisy data, compromising the dependability of AI predictions. Labelled data, which is important for supervised learning, is frequently sparse in IIoT, and labelling requires expert knowledge, which is highly time-consuming and costly.

It may be difficult for AI models that were trained on historical data to adjust to novel or undiscovered cyberthreats. Systems are left susceptible by static models that are unable to adapt to new threats in the IIoT, where cyber-physical systems are continuously subjected to changing attack techniques. Adversarial assaults make the problem worse by manipulating input data to trick AI programs. Additionally, retraining AI models to incorporate new threat patterns is often slow, requiring sufficient new data, expert oversight, and testing.

Many AI models, particularly deep learning models, function as "black boxes," making it challenging to understand how they function within. Implementing AI in IIoT also comes with high costs and technical complexities. Deploying AI often needed significant infrastructure upgrades, specialized skills in AI, cybersecurity, and industrial operations, and ongoing operational costs for detecting, maintenance, and retraining. These costs can be special for small-and medium-sized enterprises (SMEs). Overcoming the limitations, AI models for IIoT need to optimize resource constraints, improve data quality, and adapt to evolving threats. Developing lightweight models, such as TinyML, and implementing model compression techniques such as pruning and quantization can address resource issues. XAI clarifies clearly how algorithms produce assessments, recommendations, and customized learning pathways [29]. While present AI models show assurance in enhancing IIoT security, addressing problems related to resource constraints, data quality, adaptability, and interoperability is important for optimal performance. Promoting transparency, cost-efficiency, security, and resilience will be essential to fully harness AI's potential in integrating and optimizing IIoT systems, ultimately enabling more robust and trustworthy industrial operations.

13.9 OPPORTUNITIES AND IMPROVEMENTS IN INDUSTRIAL IoT SECURITY

Modern artificial intelligence (AI) developments are greatly improving the capacity to safeguard IIoT environments by means of creative near-threat detection, prevention, and response. By employing unsupervised learning techniques such as clustering, which may identify data pattern deviations without labelled data, artificial intelligence has improved anomaly detection, a crucial security feature in IIoT systems. Two deep learning models that are being used empirically to process intricate time-series data and identify subtle irregularities are convolutional and recurrent neural networks. Auto encoders enable early and precise threat detection by assisting in the reconstruction of typical data patterns and flagging deviations.

IIoT systems are becoming more capable of detecting and reacting to hazards on their own by applying reinforcement learning (RL), which uses real-time feedback. By automating reactions to abnormalities and dynamically optimizing security policies, RL models can decrease human intervention and improve security flexibility. This fosters conviction and guarantees accountability, which is particularly important in regulated sectors such as healthcare and banking. The goal of adversarial machine learning is to increase AI models' resistance to malevolent manipulation. AI systems are better able to withstand issues and carry on operating dependably because of adversarial training and the creation of more robust model architectures. Edge AI lowers latency and improves performance by processing data locally on edge devices. Graph Neural Networks (GNNs) have become a promising tool for anomaly detection in IIoT, using the relational structure of data and interactions among connected devices [30]. The integration of AI and quantum computing is anticipated to greatly enhance IIoT security. It is anticipated that the ability of quantum computing to solve complicated problems at previously unheard-of speeds will revolutionize data processing, cryptography, and hazard detection. To defend IIoT systems from quantum-enabled attacks, researchers are investigating lattice-based cryptography [31] and other quantum-resistant encryption methods. Large volumes of IIoT data can be processed more quickly by AI-powered quantum algorithms, which can speed up anomaly detection. Quantum machine learning models, including quantum support vector machines, can detect threats faster, increasing the speed and accuracy of threat response. The potential use of quantum computing for complex testing and simulations is another beneficial application in IIoT security. When FL and ATR are combined, a strong security framework for IIoT is produced that provides dynamic real-time reaction and decentralized threat intelligence sharing. These technologies offer a proactive, scalable, and privacy-conscious security solution. By identifying minute trends in massive datasets and anticipating possible threats, AI-driven security transforms IIoT defense from being reactive to proactive. Federated learning generates insights to identify threats while protecting the privacy of data. Quantum computing is going to enhance AI capabilities, provide quantum-resistant encryption, and enhance threat detection. Industries may improve operational efficiency, reduce their environmental impact, and foster sustainable growth by utilizing cutting-edge technologies such as artificial intelligence (AI), machine learning, and the Internet of Things [32]. AI will help IIoT operators work together more effectively to share threat intelligence and build robust protection networks.

13.10 CONCLUSION

AI developments are revolutionizing IIoT security by making it possible for more complex, flexible, and reliable protection systems. IIoT is being made possible by methods such as edge AI, federated learning, reinforcement learning, and enhanced anomaly detection. Settings with the necessary resources to protect data and react to attacks instantly are essential for maintaining privacy and ensuring robust security. Combining cutting-edge strategies such as explainable AI and adversarial machine learning contributes to the reliability and dependability of these systems even in intricate, crucial industrial environments. These developments are not just increasing the security of IIoT environments but are also paving the path for the safe and effective growth of IIoT applications in various sectors.

REFERENCES

[1] Anisha PR, Kishor Kumar Reddy C, Nguyen Nhu Gia, Bhushan Megha, Kumar Ashok, Hanafiah Marlia Mohd, "Intelligent Systems and Machine Learning for Industry: Advancements, Challenges, and practices" In 2022 Dec 21, CRC Press.

[2] Sadeghi AR, Wachsmann C, Waidner M. "Security and privacy challenges in industrial internet of things". In *Proceedings of the 52nd Annual Design Automation Conference* 2015 Jun 7 (pp. 1–6).

[3] Yu X, Guo H. "A survey on IIoT security." In *2019 IEEE VTS Asia Pacific Wireless Communications Symposium (APWCS)* 2019 Aug 28 (pp. 1–5). IEEE.

[4] Bakić B, Milić M, Antović I, Savić D, Stojanović T. "10 years since Stuxnet: What have we learned from this mysterious computer software worm?" In *2021 25th International Conference on Information Technology (IT)* 2021 Feb 16 (pp. 1–4). IEEE.

[5] Falliere N, Murchu LO, Chien E. W32. "Stuxnet dossier". *White paper, Symantec corp., security response.* 2011 Feb 11;5(6): 29.

[6] Al-Zahrani FS, Hassan N. "Industrial Internet of Things: A Cyber Security Perspective Investigation". In *2023 1st International Conference on Advanced Innovations in Smart Cities (ICAISC)* 2023 Jan 23 (pp. 1–6). IEEE.

[7] Jassas MS, Mahmoud QH. "Evaluation of failure analysis of IoT applications using edge- cloud architecture". In *2022 IEEE International Systems Conference (SysCon)* 2022 Apr 25 (pp. 1–8). IEEE.

[8] Liang F, Hatcher WG, Liao W, Gao W, Yu W. "Machine learning for security and the internet of things: the good, the bad, and the ugly". *IEEE Access.* 2019 Oct 22; 7:158126–158147.

[9] Jayalaxmi PL, Saha R, Kumar G, Conti M, Kim TH. "Machine and deep learning solutions for intrusion detection and prevention in IoTs: A survey". *IEEE Access.* 2022 Nov 7; 10:121173–121192.

[10] Ahmad SF, Ferjani MY, Kasliwal K. "Enhancing security in the industrial IoT sector using quantum computing". In *2021 28th IEEE International Conference on Electronics, Circuits, and Systems (ICECS)* 2021 Nov 28 (pp. 1–5). IEEE. DOI: 10.1109/ICECS53924.2021.9665527

[11] Liao H, Murah MZ, Hasan MK, Aman AH, Fang J, Hu X, Khan AU. "A survey of deep learning technologies for intrusion detection in internet of things". *IEEE Access.* 2024 Jan 2 DOI: 10.1109/ACCESS.2023.3349287

[12] Javaid M, Haleem A, Singh RP, Suman R. "Substantial capabilities of robotics in enhancing industry 4.0 implementation". *Cognitive Robotics.* 2021 Jan 1; 1:58–75 DOI: 10.1016/j.cogr.2021.06.001

[13] Awotunde JB, Chakraborty C, Adeniyi AE. "Intrusion detection in industrial internet of things network-based on deep learning model with rule-based feature selection". *Wireless Communications and Mobile Computing.* 2021;2021(1):7154587 DOI:10.1155/2021/7154587

[14] Walling S, Lodh S. "A survey on intrusion detection systems: Types, datasets, machine learning methods for NIDS and challenges". In *2022 13th International Conference on Computing Communication and Networking Technologies (ICCCNT)* 2022 Oct 3 (pp. 1–7). IEEE.

[15] A. Singh, A. Mishra, A. Antil, B. Bhushan and A. Chauhan. "Anomaly Based IDS in Industrial IoT." In *2023 International Conference on Smart Systems for applications in Electrical Sciences (ICSSES)*, Tumakuru, India, 2023, pp. 1–6, DOI: 10.1109/ICSSES58299.2023.10199661

[16] Kumar GK, Kumar RR, Reddy KN, Babu PA. "Ensemble of support vector machines using fuzzy-pam for intrusion detection". *J. Mech. Cont.& Math. Sci.*, Special Issue, No.-5, pp 109–119, 2020, DOI:10.26782/jmcms.spl.5/2020.01.00009

[17] Al-Emadi S, Al-Mohannadi A, Al-Senaid F. "Using deep learning techniques for network intrusion detection". In *2020 IEEE international conference on informatics, IoT, and enabling technologies (ICIoT)* 2020 Feb 2 (pp. 171–176). IEEE DOI: 10.1109/ICIoT48696.2020.9089524

[18] Vinh TQ, Huy NT." Predictive maintenance IOT system for industrial machines using random forest regressor". In *2022 International Conference on Advanced Computing and Analytics (ACOMPA)* 2022 Nov 21 (pp. 86–91). IEEE DOI: 10.1109/ACOMPA57018.2022.00020
[19] Alampalayam SP, Kumar A. "Predictive security model using data mining". In *IEEE Global Telecommunications Conference, 2004. GLOBECOM'04*. 2004 Nov 29 (Vol. 4, pp. 2208–2212). IEEE. DOI: 10.1109/GLOCOM.2004.1378401
[20] Al-Hawawreh M, Moustafa N, Garg S, Hossain MS. "Deep learning-enabled threat intelligence scheme in the internet of things networks". *IEEE Transactions on Network Science and Engineering*. 2020 Oct 20;8(4):2968–2981. DOI:10.1109/TNSE.2020.3032415
[21] Hui H, Zhou C, Xu S, Lin F. "A novel secure data transmission scheme in industrial internet of things". *China Communications*. 2020 Jan 28;17(1):73–88.
[22] Balusa VS, Srinivas K. "An effective intrusion detection system for edge computing using ConvNeXt and ResNet152V2." *International Journal of Computational Intelligence and Applications*. 2024 Apr 25:2450014 DOI:10.1142/S1469026824500147
[23] El-Gendy S. "IoT based AI and its implementations in industries". In *2020 15th International Conference on Computer Engineering and Systems (ICCES)* 2020 Dec 15 (pp. 1–6). IEEE.
[24] Ni C, Li SC. "Machine learning enabled industrial IoT security: Challenges, trends and solutions". *Journal of Industrial Information Integration*. 2024 Jan 23:100549.
[25] Tatineni S. "Ethical considerations in AI and data science: Bias, fairness, and accountability". *International Journal of Information Technology and Management Information Systems (IJITMIS)*. 2019;10(1):11–21.
[26] Venkatasubbu S, Krishnamoorthy G. "Ethical considerations in AI addressing bias and fairness in machine learning models." *Journal of Knowledge Learning and Science Technology* ISSN: 2959-6386 (online). 2022 Sep 14;1(1):130–138.
[27] Kishor Kumar Reddy C, Anisha PR, Khan Samiya, Hanafiah Marlia Mohd, Lavanya Pamulaparty R Mohana Madana. "Challenges and Future Prospects In The Transition To Industry" In *Sustainability in Industry: Theory and Applications*, 2024 Feb 19, CRC Press.
[28] Sinha S, Lee YM. "Challenges with developing and deploying AI models and applications in industrial systems." *Discover Artificial Intelligence*. 2024 Aug 16;4(1):55.
[29] Fatima S, Reddy CK, Sunerah A, Doss S. "Innovations in Education: Integrating Explainable AI into Educational Intelligence." In *Internet of Behaviour-Based Computational Intelligence for Smart Education Systems* 2025 (pp. 19–52). IGI Global.
[30] Wu Y, Dai HN, Tang H. "Graph neural networks for anomaly detection in industrial Internet of Things". *IEEE Internet of Things Journal*. 2021 Jul 2; 9(12):9214–9231.
[31] Pradhan PK, Rakshit S, Datta S. "Lattice based cryptography: Its applications, areas of interest & future scope". In *2019 3rd International Conference on Computing Methodologies and Communication (ICCMC)* 2019 Mar 27 (pp. 988–993). IEEE. DOI: 10.1109/ICCMC.2019.8819706
[32] Reddy CKK, Anisha PR, Hanafah MM, Doss S and Lipert KJ, 2024. "Intelligent systems and industrial internet of things for sustainable development." ISBN 9781032640914.

Index

Pages in **bold** refer to tables.

For Product Safety Concerns and Information please contact our EU representative GPSR@taylorandfrancis.com Taylor & Francis Verlag GmbH, Kaufingerstraße 24, 80331 München, Germany